110 Semiconductor Projects for the Home Constructor

Other Books by R. M. Marston

20 Solid State Projects for the Home
20 Solid State Projects for the Car and Garage
110 Integrated Circuit Projects for the Home Constructor
110 Thyristor Projects using S.C.R.S. and Triacs

110 Semiconductor Projects for the Home Constructor

R. M. MARSTON

LONDON
ILIFFE BOOKS

THE BUTTERWORTH GROUP

ENGLAND

Butterworth & Co (Publishers) Ltd
London: 88 Kingsway, WC2B 6AB

AUSTRALIA

Butterworth Pty Ltd
Sydney: 586 Pacific Highway, NSW 2067
Melbourne: 343 Little Collins Street, 3000
Brisbane: 240 Queen Street, 4000

CANADA

Butterworth & Co (Canada) Ltd
Toronto: 14 Curity Avenue, 374

NEW ZEALAND

Butterworths of New Zealand Ltd
Wellington: 26-28 Waring Taylor Street, 1

SOUTH AFRICA

Butterworth & Co (South Africa) (Pty) Ltd
Durban: 152-154 Gale Street

First published in 1969 by
Iliffe Books, an imprint
of the Butterworth Group

Second impression 1971
Third impression 1973

ISBN 0 592 02845 3 Standard
0 592 02864 X Limp

Printed in England by Hazell Watson & Viney Ltd,
Aylesbury, Bucks

CONTENTS

PREFACE

Semiconductor technology has advanced so rapidly in the past decade that many amateurs, technicians, and engineers have found great difficulty in keeping track of the new devices that have become available. Consequently, many outstandingly useful devices, like the field-effect transistor, the unijunction transistor, the silicon controlled-rectifier, and the integrated circuit, have remained unused by many amateurs and professionals.

This 'usage gap' is due mainly to the lack of readable information on the new devices. Most books and articles that deal with them get bogged down in a morass of useless theory and incomprehensible mathematics. This present volume manages to overcome this problem. It sets out to introduce the reader to the new devices by experiment, rather than theory. Each chapter starts by outlining the basic characteristics of a new device, rather than its intricate theory, and then goes on to give a range of practical circuits in which it is used. 110 different circuits are described, and the operation of each one is explained in simple and concise terms.

The volume is intended to appeal equally to the amateur and professional electronics man. The explanations of device operation are meant to be readable by the amateur with no mathematical knowledge, while at the same time conveying information of value to the technician and engineer. The practical circuits should be of interest to all readers. Those of particular interest to the amateur include simple amplifiers, lamp and relay driving circuits, electronic switches that can be operated by light, by sound, or by contact with water, and electronic timer and delay circuits giving periods ranging from a fraction of a second to 35 min.

Circuits of particular interest to the technician and engineer include amplifiers with input impedances as high as 500 MΩ, voltage and current regulators, a constant-volume amplifier, pulse and other waveform generators, analogue-to-digital converters, logic circuits, frequency dividers, a d.c. chopper, and simple power controller circuits. All circuits are designed around internationally available semiconductors, so the parts needed in all construction projects should be readily obtainable in all parts of the western world.

R. M. Marston

CHAPTER 1

30 SILICON-PLANAR TRANSISTOR PROJECTS

Recent years have seen many advances in semiconductor production techniques. Amongst the most important of these have been the introduction of simplified methods of manufacturing silicon-planar multi-junction networks, and the widespread adoption of epoxy or plastic encapsulation techniques. The combination of these techniques has resulted in a new generation of low-cost high-performance transistors, having many advantages over the earlier germanium types. These new transistors have very low leakage currents, are capable of operating at high temperatures, and can withstand considerable physical and electrical abuse without breaking down.

With these advantages in mind, let's take a look at the characterisitcs of just two low-cost general-purpose silicon-planar transistors, and then go on to consider thirty or so useful little circuits in which they can be used.

The two transistors that we'll select for this purpose are the 2N2926 npn type by G.E.C., and the 2N3702 pnp type by Texas. Their general characteristics and lead connections are shown in Fig. 1.1 and Table 1.1. Note that the 2N2926 type is colour coded according to gain; we'll use the medium-gain 'orange' type in most applications.

Using silicon-planar transistors

The most striking differences between silicon and germanium, and pnp and npn transistors are shown in Fig. 1.2. Although no component values are shown here, typical circuit potentials are included,

Table 1.1
GENERAL CHARACTERISTICS OF THE
2N2926 AND 2N3702 TRANSISTORS

	2N2926	*2N3702*
Transistor Type	npn	pnp
I_c (max)	100 mA	200 mA
V_{ceo} (max)	18 V	25 V
V_{cbo} (max)	18 V	40 V
f_T (min) = gain/bandwidth product	120 MHz	100 MHz
h_{fe} (= a.c. beta)	55-100 at 2 mA (code red) 90-180 at 2 mA (code orange) 150-300 at 2 mA (code yellow) 235-470 at 2 mA (code green)	60-300 at 50 mA
I_{cbo} (max)	0.5 μA	0.1 μA
P_{tot} (max)	200 mW	300 mW

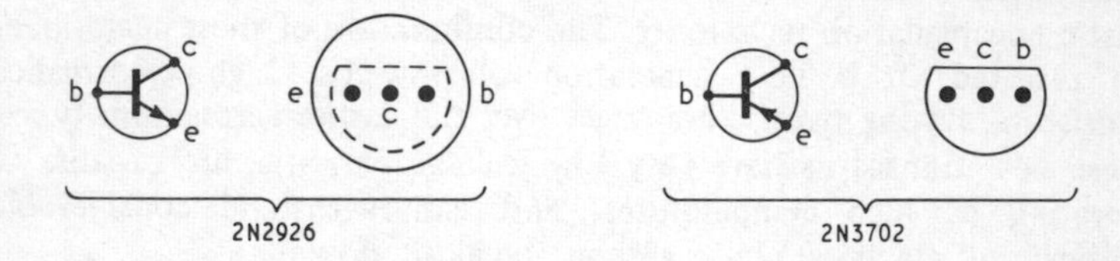

Fig. 1.1
Symbols, and lead connections (looking into the base) of the 2N2926 and 2N3702 transistors

and the most important point to notice is that the emitter-base potentials of the silicon transistors are 0.65 V, while that of the germanium is only 0.2 V. This difference between the emitter-base junction potentials is the most significant point to bear in mind when designing amplifiers that are in other ways similar. In the case of Fig. 1.2, the germanium pnp circuit can be modified to operate with a silicon transistor by simply altering the value of R_1 to give the required base potential, leaving R_2, R_3, and R_4 unaltered. It can be made to work with an npn silicon type by also transposing the supply connections, as in Fig. 1.2c.

Although conventional germanium transistor circuits can be easily arranged to work with silicon types, such an approach is rather pointless, since it does not take full advantage of the benefits offered by silicon transistors. With this point in mind, some practical circuits will now be considered.

Simple common emitter amplifiers

As shown in Fig. 1.2a, germanium transistors require fairly complex biasing networks; R_1, R_2, R_4, and C_2 are used for this purpose. This complexity is needed partly to allow for differences in the current gains of individual transistors, but mainly to compensate for the large leakage currents that are inherent with germanium transistors. Silicon transistors, on the other hand, have very low leakage currents, and

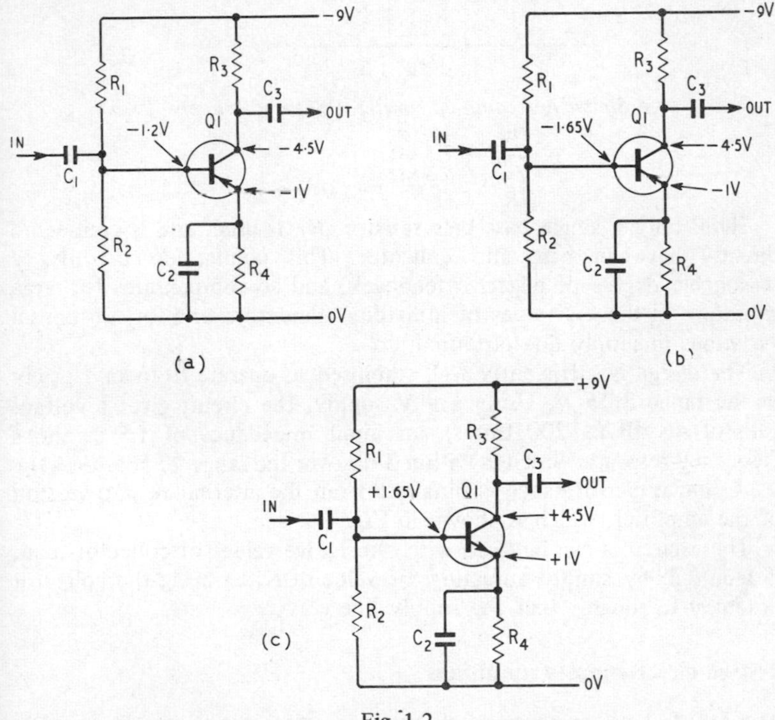

Fig. 1.2

Similar common emitter circuits, using different types of transistor. Note the differences between the emitter-base potentials of germanium and silicon transistors, and the differences in supply polarity of npn and pnp types.
(a) pnp germanium circuit (b) pnp silicon circuit (c) npn silicon circuit

their bias networks can thus be considerably simplified, with no deterioration in performance. Fig. 1.3 shows a simple common emitter amplifier designed around an npn silicon-planar transistor.

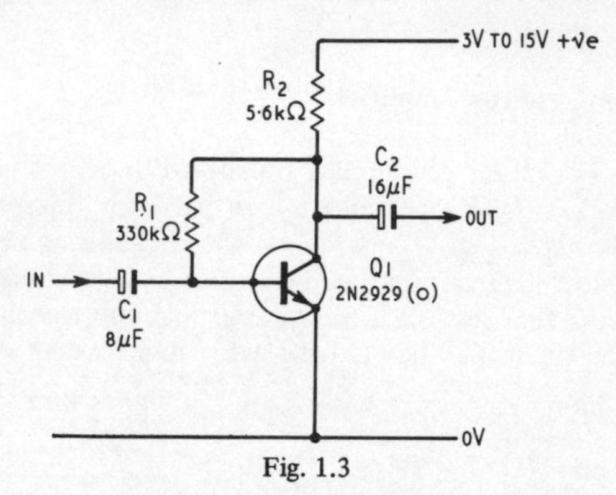

Fig. 1.3

Simple npn common emitter. Using a 9 V supply:

A_V = 46 dB
Z_{in} = 1.5 kΩ
Z_{out} = 5.6 kΩ
f_R = 27 Hz–120 kHz ± 3 dB

Here, only a single base-bias resistor, R_1, is used, and is connected directly between base and collector. This connection provides a reasonable degree of negative feedback, and so compensates for large variations in the h_{fe} values of individual transistors, and for substantial variations in supply line potential.

The design is sufficiently well stabilised to operate from any supply in the range 3–15 V. Using a 9 V supply, the circuit gives a voltage gain of 46 dB (= 200 times), an input impedance of 1.5 k, and a frequency response which is within 3 dB over the range 27 Hz–120 kHz.

A similar performance is obtained from the alternative pnp version of the amplifier, which is shown in Fig. 1.4.

These circuits can be used with alternative values of collector load, if required, by simply adjusting the value of R_1 to bring the collector potential to roughly half the supply line voltage.

2-Stage direct coupled amplifiers

The low leakage currents of silicon transistors enable direct coupling to be used between amplifier stages in many applications, and Fig. 1.5 shows a typical 2-stage direct coupled circuit designed around npn silicon transistors.

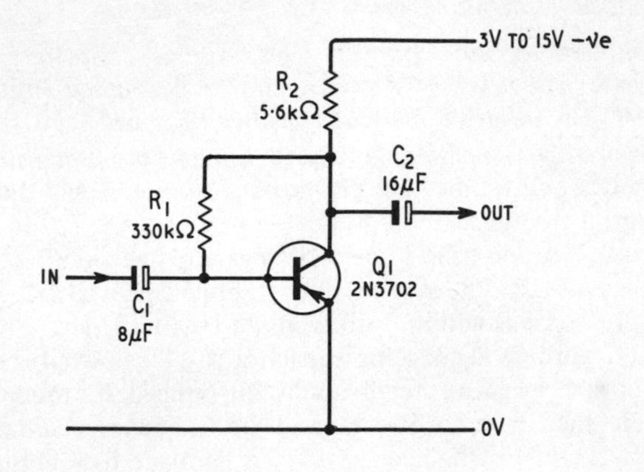

Fig. 1.4

Simple pnp common emitter amplifier. Performance is similar to that of Fig. 1.3

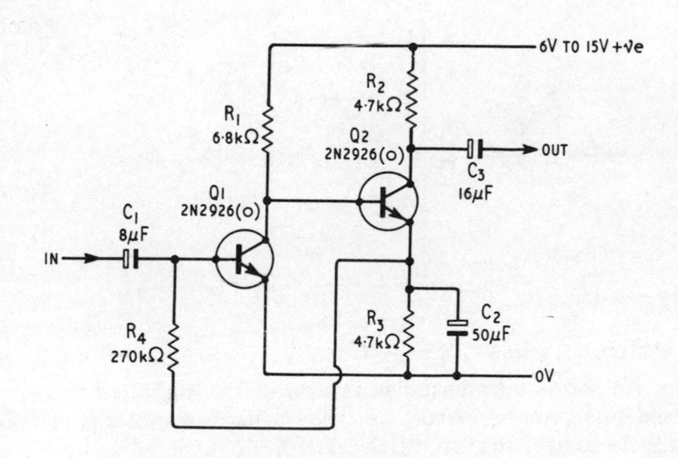

Fig. 1.5

2-stage direct coupled amplifier. Using a 9 V supply:

A_V = *76 dB*
Z_{in} = *3.9 kΩ*
Z_{out} = *4.7 kΩ*
f_R = *35 Hz–35kHz ± 3 dB*

Both transistors are connected as common emitter amplifiers, and the base-bias of *Q*1 is derived from the decoupled emitter of *Q*2. Substantial d.c. negative feedback is thus obtained, and the circuit's working potentials are well stabilised against variations in transistor characteristics and supply line potential. The circuit will operate from any supply in the range 6–15 V.

Using a 9 V supply, the total voltage gain of the circuit is 76 dB, the input impedance is 3.9 k, the output impedance is 4.7 k, and the frequency response is within 3 dB from 35 Hz to 35 kHz.

If the *Q*2 emitter decoupling capacitor, C_2, is removed, a substantial amount of a.c. negative feedback is introduced to the circuit; the voltage gain then falls to 46 dB, and the frequency response extends from 35 Hz to 120 kHz. The circuit can be made to give intermediate values of gain and frequency response, if required, by replacing R_3 with a 5 kΩ pot, and connecting C_2 between its slider and ground.

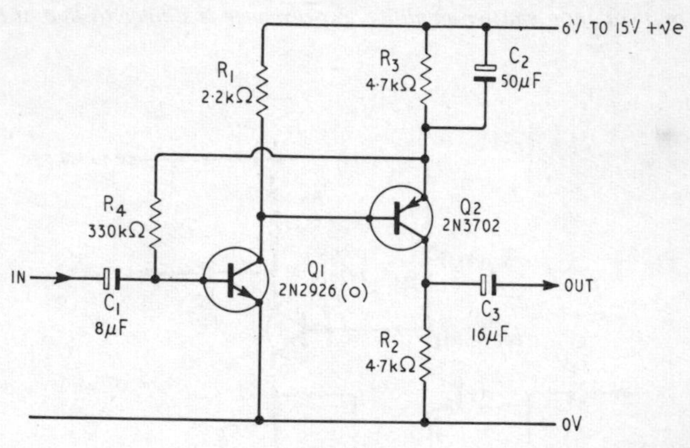

Fig. 1.6

Alternative 2-stage amplifier. Performance is similar to that of Fig. 1.5

Fig. 1.6 shows an alternative version of the amplifier. It uses one npn and one pnp transistor, but gives a performance that is almost identical to that of the circuit of Fig. 1.5.

The two amplifiers shown in Figs. 1.5 and 1.6 each give an output at *Q*2 collector that is in phase with, but much greater than, the input signal at *Q*1 base. Consequently, any signal feedback that occurs between the output and the input will be regenerative, so the amplifiers may tend to be unstable if the supply lines are not properly decoupled, or if the input connections are not screened. This snag is overcome in the circuit of Fig. 1.7.

Here, $Q1$ is wired as a common emitter amplifier, and has its collector directly coupled to the base of emitter follower $Q2$. 180° of signal phase shift naturally occurs between the base and collector of the $Q1$ stage, but zero phase shift occurs between the base and emitter of $Q2$, so a total of only 180° phase shift occurs between the input at $Q1$ base and the output at $Q2$ emitter, and any feedback that occurs is degenerative. $Q1$ base-bias is derived from $Q2$ emitter via R_1, so negative feedback biasing is used, and the circuit's working potentials are well stabilised.

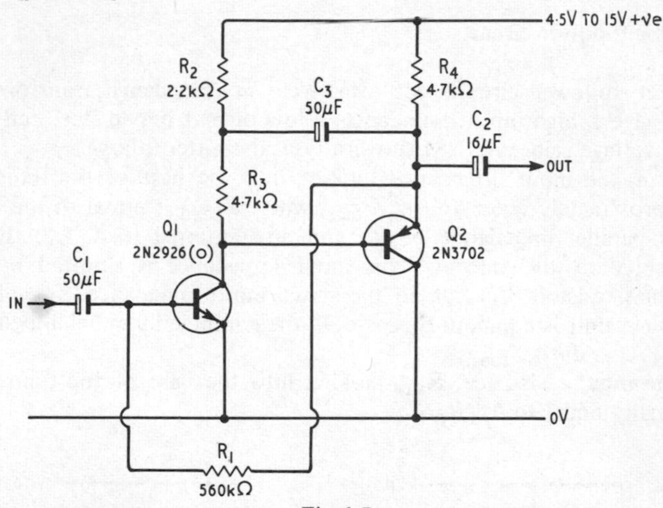

Fig. 1.7

Direct coupled amplifier with bootstrapped common emitter stage. Using a 9 V supply and an input signal from a 1 kΩ source:

A_V = 66 dB
Z_{in} = 330 Ω
Z_{out} = 820 Ω
f_R = 20 Hz–32 kHz ± 3 dB

Now, the signal appearing at $Q2$ emitter is almost identical with that at $Q1$ collector, but is at a low impedance and is effectively isolated from it. In Fig. 1.7, this low impedance signal is fed, via C_3, to the junction of the R_2-R_3 split collector load of $Q1$. Consequently, almost identical a.c. signals appear at both ends of R_3, and only a negligible signal current flows in this resistor, which thus appears as a very high impedance to a.c. signals; the effective a.c. value of R_3 is in fact increased to several hundred kilohms by the use of this feedback or 'bootstrap' technique, and $Q1$ therefore gives a very high voltage gain, which is finally made available at the emitter of $Q2$ at a fairly low impedance level.

This circuit will operate from any supply in the range 4.5–15 V. Using a 9 V supply, it gives a voltage gain of about 66 dB, an input impedance of 330 Ω, and an output impedance of 820 Ω. The frequency response varies somewhat with the source impedance of the input signal; with a 100 Ω source, the 3 dB points occur at 30 Hz and 45 kHz, and with a 1 kΩ source at 20 Hz and 32 kHz.

Emitter follower circuits

Emitter follower circuits act effectively as impedance transformers. They give a high input impedance, a low output impedance, and near unity voltage gain. Fig. 1.8a shows a typical emitter follower.

Here, the input impedance looking into the base of the transistor is approximately equal to $h_{fe}.Z_{load}$, where Z_{load} is equal to the combined parallel impedance of R_e and any external load, Z_x, that is connected at the output. This input impedance is shunted by the base-bias resistors $(R_1$-$R_2)$, so the actual input impedance, Z_{in}, of the complete unit is equal, in this case, to the combined parallel impedance of R_1, R_2 and $h_{fe}.Z_{load}$.

The input resistance, R_{in}, looking into the base of the transistor, is roughly equal to $h_{fe}R_e$.

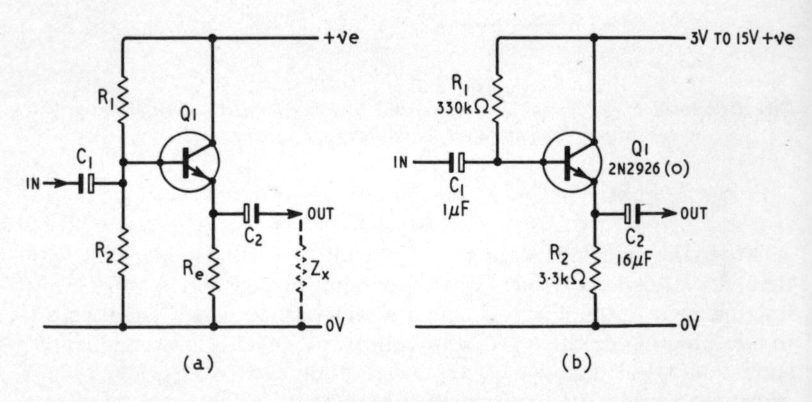

Fig. 1.8

(a) Typical emitter follower circuit (see text) (b) Simple emitter follower giving a Z_{in} *of 180 kΩ*

To enable the emitter follower to handle the largest possible signal levels, it is usually biased so that its emitter is at a quiescent potential of roughly half the supply line voltage. The standard way of achieving this in germanium circuits, where base leakage currents are large and may be comparable to normal bias currents, is to wire R_1 and R_2 as a potential divider network, as in the diagram. The emitter of a transistor inevitably takes up a potential that is within a fraction of a volt of that on its base, so, if $R_1 = R_2$, and R_2 is small relative to R_{in}, the required bias conditions are naturally met, and are not greatly altered by normal variations in the leakage currents of germanium transistors. The major snag with this method of biasing is that the bias resistors impose a severe restriction on the maximum available input impedance of the circuit.

Silicon transistors, on the other hand, have very low leakage currents, so, assuming that these are low relative to the normal base-bias currents, the required bias conditions can be met by simply wiring a single resistor, R_1, with a value equal to R_{in}, between the base of $Q1$ and the +ve supply line, as in the practical circuit of Fig. 1.8b. R_1 and R_{in} then act effectively as a potential divider base-bias network, setting $Q1$ base and emitter at roughly half of the supply line voltage, but cause only a small reduction in the available input impedance of the circuit.

Using the component values shown, the circuit of Fig. 1.8b can be used with any supply in the range 3–15 V, and gives an input impedance, with the output unloaded, of about 180 kΩ at all voltages. Alternative values of Z_{in} can be obtained by changing the values of R_1 and R_2. R_1 should have a value of roughly $100 \times R_2$; the values should be chosen so that R_2 draws a quiescent current within the limits 0.5 mA–20 mA.

If input impedances substantially greater than a couple of hundred kilohms are required, the circuit of Fig. 1.9 can be used. Here, $Q1$ and $Q2$ are wired in the Darlington or super-alpha mode, with the emitter current of $Q1$ feeding directly into the base of $Q2$, and act like a single transistor with a gain roughly equal to the product of the two individual h_{fe} values. In this mode, $Q1$ operates at such a low current level that leakage currents become significant; to minimise the effects of these, R_4 is used as a stabilising resistor, and base biasing is provided by voltage divider network R_1-R_2. To minimise the shunting effects of R_1 and R_2 on Z_{in}, isolating resistor R_3 is wired in place as shown, and is bootstrapped from $Q2$ emitter bia C_2.

This circuit gives an input impedance of about 3.3 MΩ. The input impedance can be reduced, if required, by lowering the value of R_4, down to a minimum of 18 kΩ, at which point Z_{in} = 1 MΩ. Alternatively, the input impedance can be raised to about 5 MΩ, by using a green coded 2N2926 transistor in the $Q1$ position.

An alternative way of obtaining a very high input impedance and near unity voltage gain is shown in Fig. 1.10. In this circuit, $Q1$ and $Q2$ both act as common emitter amplifiers, but all of the $Q1$ collector signal current flows directly into the base of $Q2$, and all of the $Q2$ signal current flows through R_3; thus, the R_3 signal current is roughly equal to the $Q1$ base current times the product of the individual

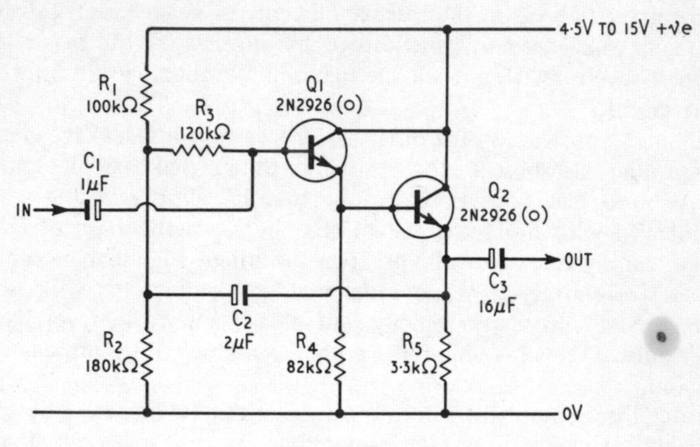

Fig. 1.9
Bootstrapped 2-stage emitter follower giving a Z_{in} of 3.3 MΩ

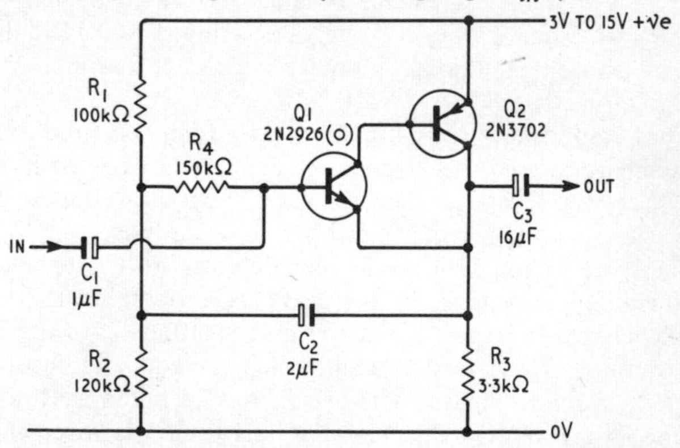

Fig. 1.10
Complementary feedback pair circuit, giving a Z_{in} of 6 MΩ

transistor gains, and the input impedance to the base of $Q1$ is roughly equal to $R_3 h_{fe1} h_{fe2}$. As far as voltage gains are concerned, virtually 100% negative feedback is obtained overall, so the circuit gives a gain of almost exactly unity. Thus, the circuit of Fig. 1.10, which is known as a complementary feedback pair, gives a performance very similar to that of a 2-stage emitter follower.

R_1 and R_2 form a voltage divider base-bias network, which is effectively isolated from $Q1$ base by bootstrapped resistor R_4. The circuit can be used with any supply in the range 3–15 V, and gives an input impedance of about 6 MΩ. This impedance can be raised to about 10 MΩ, if required, by using a green coded 2N2926 transistor in the $Q1$ position.

Relay operating circuits

Transistors can be used to modify the characteristics of simple and inexpensive relays, either to effectively increase their current or voltage sensitivities, or to give them a built-in operating time delay.

Fig. 1.11a shows a simple circuit in which $Q1$ is wired as an emitter follower and uses a relay as its emitter load, thus effectively increasing the relay's current sensitivity by about 50 times. R_2 shunts base leakage currents to ground in the absence of an input bias, and should have a value 100 times greater than the relay's coil resistance. R_1 limits the base current to a safe value in the event of an excessive operating voltage being connected at the input. $D1$ prevents any back e.m.f. from damaging the circuit as the relay switches rapidly on or off.

The actual relay used in this circuit (and all others described in this section) can be any type requiring an operating current less than 50 mA, and needing an operating potential less than 15 V. The circuit's supply rail should be at least 3 V greater than the operating voltage of the relay.

For correct operation of Fig. 1.11a, the input voltage must be connected with the polarity shown in the diagram. For some purposes, however, it may be required that the relay be operated with either polarity of input, and this can be achieved by wiring a bridge rectifier in the input, as shown in Fig. 1.11b. Diodes $D2$–$D5$ can be any general purpose germanium or silicon types. The input signal must, of course, be 'floating' relative to the ground line if this modification is used.

If a greater increase than fifty is needed in the relay's current sensitivity, the circuit of Fig. 1.12a can be used. Here, R_3 is given a value roughly 100 times greater than R_2 up to a maximum value of 1 MΩ, and the circuit gives an increase in current sensitivity of about

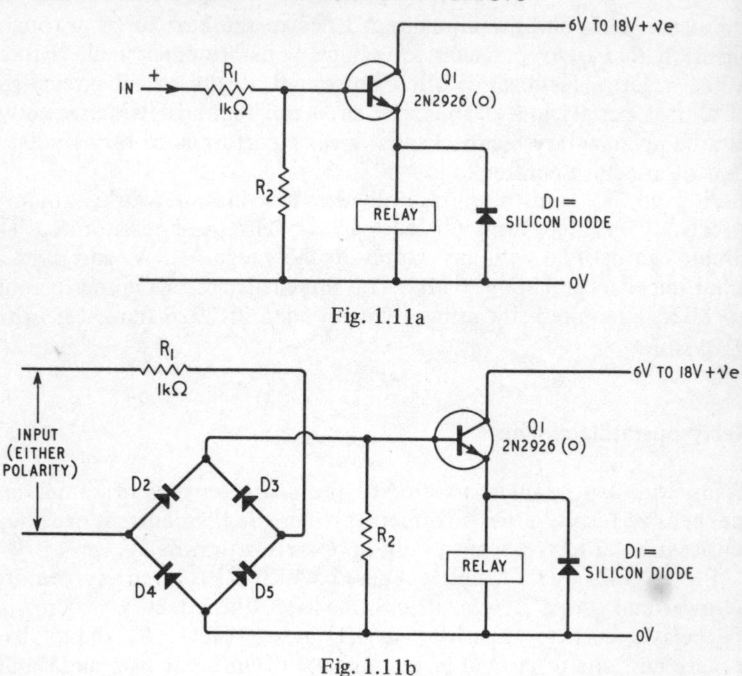

Fig. 1.11a

Fig. 1.11b

(a) Circuit for increasing relay current sensitivity by 50 times. R_2 *= 100 times relay coil resistance. (b) Modification of Fig. 1.11a for operation by either polarity input.* R_2 *= 100 times relay coil resistance,* D2–D5 *are general purpose silicon or germanium diodes*

500 times. Fig. 1.12b shows the modification for operating with either polarity of input voltage.

If an increase in both the voltage and the current sensitivity of the relay is required, the circuit of Fig. 1.13a can be used. Here, both *Q*1 and *Q*2 are wired as common emitter amplifiers. With no input connected, *Q*1 is held at cut-off by R_2, and *Q*2 is held cut-off by R_3, so the relay does not operate and the circuit consumes only a small leakage current. When an input is connected to *Q*1 base, both *Q*1 and *Q*2 are driven to saturation, and the relay operates. An input of roughly 700 mV at 40 μA is needed to drive the relay on.

Fig. 1.13b shows the modification needed for operating with either polarity of input voltage. The bridge rectifier causes some loss in the voltage sensitivity of the circuit. If *D*2–*D*5 are germanium types, the circuit needs an input of about 1.1 V to operate the relay, and if *D*2–*D*5 are silicon types, an input of nearly 2 V is needed.

Fig. 1.14 shows two circuits for imposing time delays on the operation of the relay. Fig. 1.14a gives a delay between the moment of connecting the supply and the moment at which the relay actually turns on: Fig. 1.14b causes the relay to switch on as soon as the supply is connected, but to switch off again automatically after a predetermined period. Timing periods up to about one minute are obtainable.

In Fig. 1.14a, *Q*1 and *Q*2 are wired as a Darlington emitter follower, with the base-bias of *Q*1 provided by the R_1–C_1 'potential divider' network. At the moment that the supply is first connected, C_1 is discharged and *Q*1 base is held at ground potential, so the relay is off.

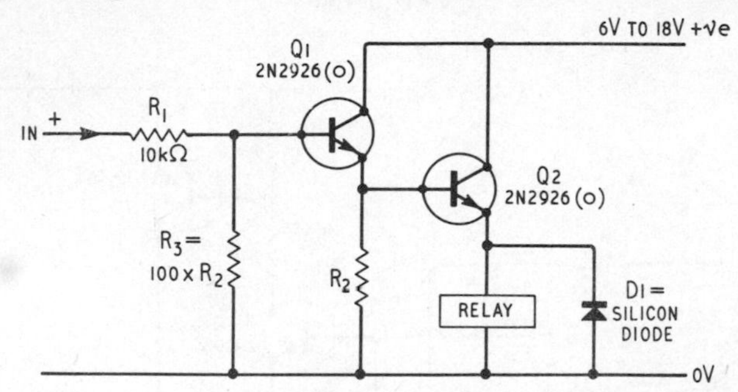

Fig. 1.12a

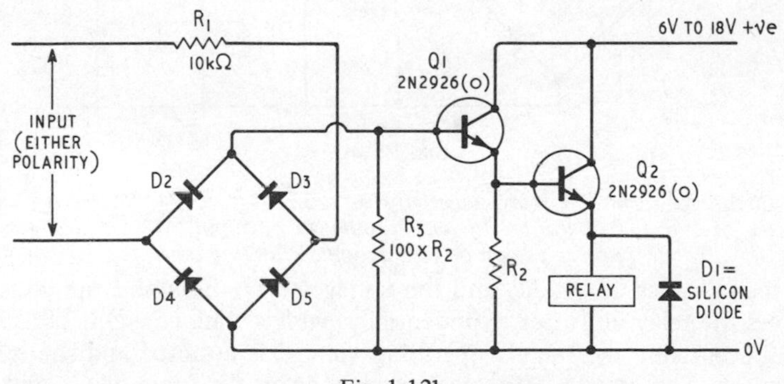

Fig. 1.12b

(a) Circuit for increasing relay current sensitivity by 50 times. R_2 = *100 times relay coil resistance. (b) Modification of Fig. 1.12a for operation by either polarity input.* R_2 = *100 times relay coil resistance,* D2–D5 *are general purpose silicon or germanium diodes*

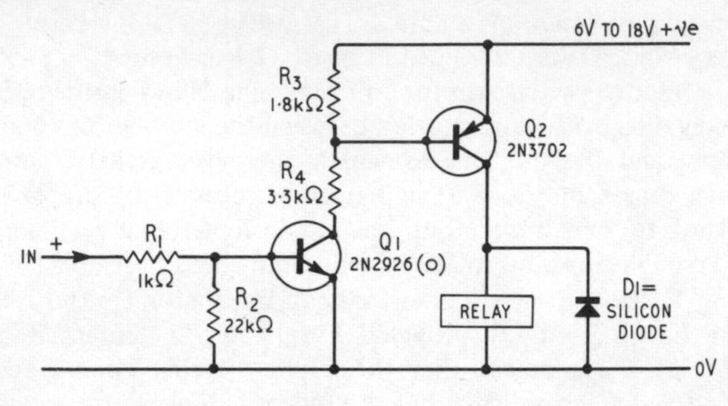

Fig. 1.13a

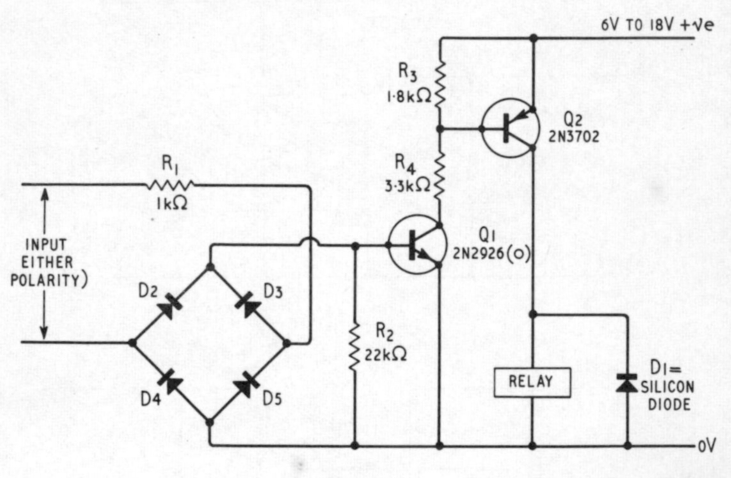

Fig. 1.13b

(a) Circuit for increasing relay sensitivity to 700 mV at 40 µA. (b) Modification of Fig. 1.13a for operation by either polarity of input. D2–D5 *are general purpose silicon or germanium diodes (see text)*

C_1 then charges up via R_1, and the voltage on $Q1$ base and the voltage across the relay coil rises exponentially, with a time constant of C_1R_1, until eventually the relay's operating voltage is attained and the relay turns on. The precise delay period depends on the value of C_1, on the relay's operating characteristics, and on the supply line potential used, but if the supply is made about 3 V greater than the relay operating voltage the delay is roughly equal to 0.1 sec/μF of C_1 value, i.e., if C_1 = 100 μF, delay = 10 sec.

*Q*1 and *Q*2 are also wired as a Darlington emitter follower in Fig. 1.14b, but in this case the positions of R_1 and C_1 are reversed. Consequently, when the supply is first connected, C_1 is discharged and *Q*1 base is shorted to the +ve supply rail, so the relay is driven hard on. C_1 then charges up via R_1, so the voltage across the relay coil decays exponentially with a time constant of $R_1.C_1$, until eventually the relay's turn-off voltage is reached. The time delay depends a great deal on the

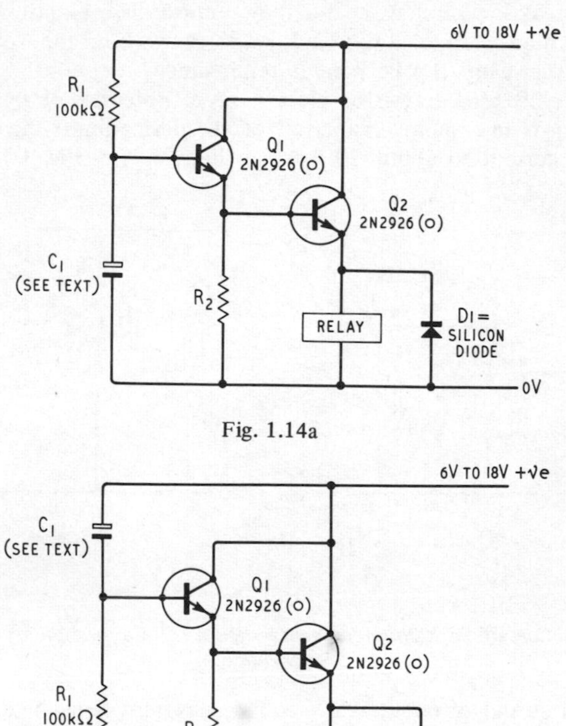

Fig. 1.14a

Fig. 1.14b

(a) Circuit for giving a switch-on delay to a relay. R_2 *= 100 times relay coil resistance. (b) Circuit giving automatic turn-off of a relay after a predetermined period.* R_2 *= 100 times relay coil resistance*

relay's on/off voltage ratio, but can be varied by choice of the C_1 value, which is thus best found by trial and error to suit individual needs.

Voltage regulator circuits

Most silicon-planar transistors have very sharply defined emitter-base reverse breakdown voltages, and their emitter-base junctions thus act as zener diodes. Figs. 1.15a and 1.15b show how the 2N2926 and 2N3702 transistors can be used as zener diodes.

The 2N2926(0) transistor gives a zener potential of 9–10 V, and the 200 mW maximum dissipation of the device limits the maximum available current to about 20 mA, so the circuit of Fig. 1.15a gives a

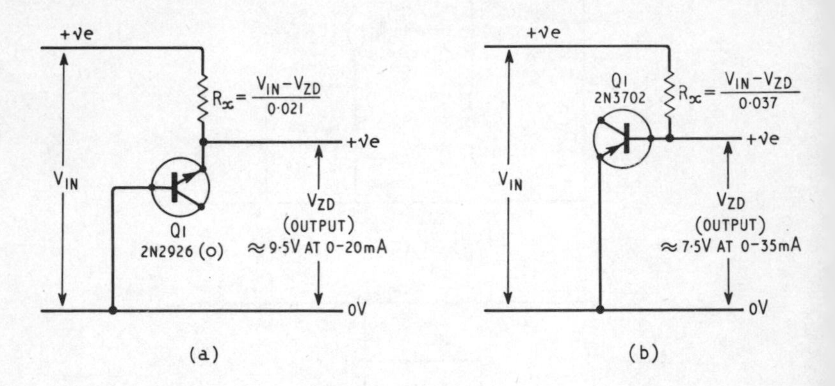

Fig. 1.15

(a) Connection of 2N2926(O) as zener diode. (b) Connection of 2N3702 as zener diode

regulated output of about 9.5 V over the current range 0–20 mA The value of V_{in} is not critical, and R_x is given by the formula in the diagram.

The 2N3702 transistor gives a zener potential of 7-8 V, and can handle maximum currents of about 37 mA. Fig. 1.15b shows a circuit giving a regulated output of about 7.5 V over the range 0–35 mA.

In both of these circuits, the R_x value is chosen to limit the zener current to the maximum permissible value, with the output unloaded.

Regulated outputs greater than 10 V can be obtained by wiring zener diodes in series. Fig. 1.16a shows how to wire two 2N2926(0) zener diodes to give an output of about 19 V at 0-20 mA, and Fig. 1.16b shows how to wire a 2N2926(0) and 2N3702 in series for an output of about 17 V at 0–20 mA.

Larger output currents can be obtained by wiring an emitter follower

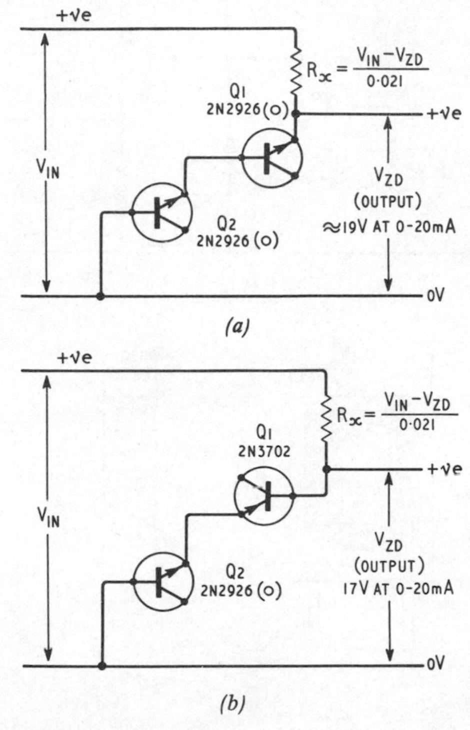

Fig. 1.16

(a) Two 2N2926(O) zener diodes wired in series to give 19 V output
(b) 2N2926(O) and 2N3702 zener diodes wired in series to give 17 V output

between the zener diodes and the output, and Fig. 1.17a shows a practical circuit giving a regulated output of about 18 V at 0–500 mA. C_1 suppresses any ripple from the unregulated line, and so gives a well smoothed output. Approximately 0.65–1.0 V are 'lost' in the emitter-base junction of *Q*3, so the regulated output is this amount less than the actual zener voltage.

*Q*3 is an MJE520 miniature silicon npn power transistor by Motorola; this transistor is complementary to the MJE370 pnp type, and Fig. 1.17b and Table 1.2 show the characteristics and connections of both types. Alternative silicon transistors can be used in the *Q*3 position if preferred,

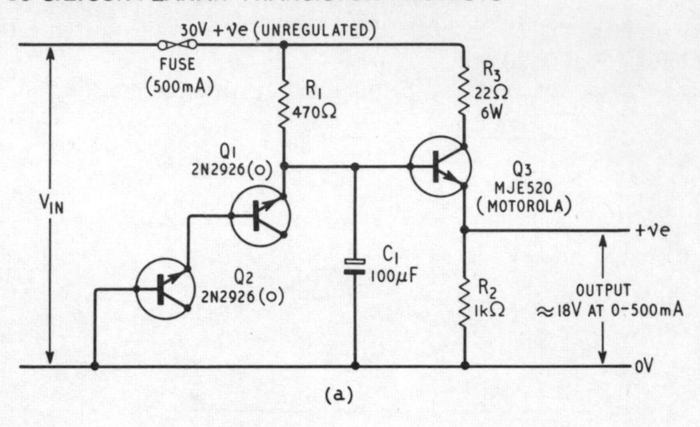

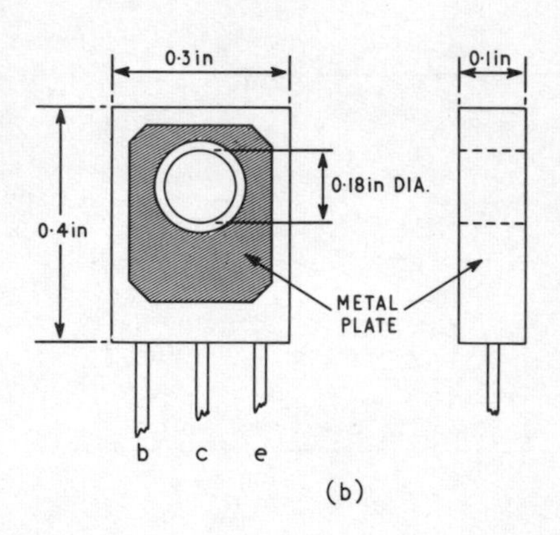

Fig. 1.17

(a) Practical 18 V regulator. (b) Dimensions and connections of the MJE520 and MJE370 miniature complementary power transistors by Motorola

but must have h_{fe} values of at least 30. The transistor has to dissipate a maximum power of about 2 W, and should be mounted on an aluminium heat sink with an area of 2 in^2.

Fig. 1.18 shows the circuit of a simple variable-voltage regulator, covering the approximate range 0–17.5 V at 0–1 A. R_2 is wired across the zener network, making a variable reference potential of 0–19 V available to the base of $Q3$; $Q3$ and $Q4$ are wired as a Darlington emitter follower, so this variable potential is made available at a high current level at $Q4$ emitter; about 1.5 V are 'lost' in $Q3$ and $Q4$,

Table 1.2

CHARACTERISTICS OF THE MJE520 AND THE MJE370 MINIATURE COMPLEMENTARY POWER TRANSISTORS BY MOTOROLA

	MJE520	*MJE370*
Transistor Type	npn	pnp
I_c (max)	3 A	3 A
V_{ceo} (max)	+40 V	–40 V
V_{cbo} (max)	+60 V	–60 V
f_T at V_{ce} 20 V	2.8 MHz	2.8 MHz
h_{FE} at I_c 0.75 A.	45–60	45–60
I_{cbo} (typ)	0.1 μA	0.1 μA
P_{tot} (max) at 45°C (on heat sink with an area of 12 in^2)	25 W	25 W

however, so the output voltage is this amount less than the zener reference potential.

In this circuit, *Q*4 may dissipate a maximum power of about 20 W, and must be mounted on a heat sink with an area of at least 12 in^2. *Q*3 dissipates a maximum power of less than 1 W, and can be mounted on a heat sink of 2 in^2. Note that R_5 must have a power rating of at least 12 W.

Note that, in the circuits of Figs. 1.17 and 1.18, the unregulated

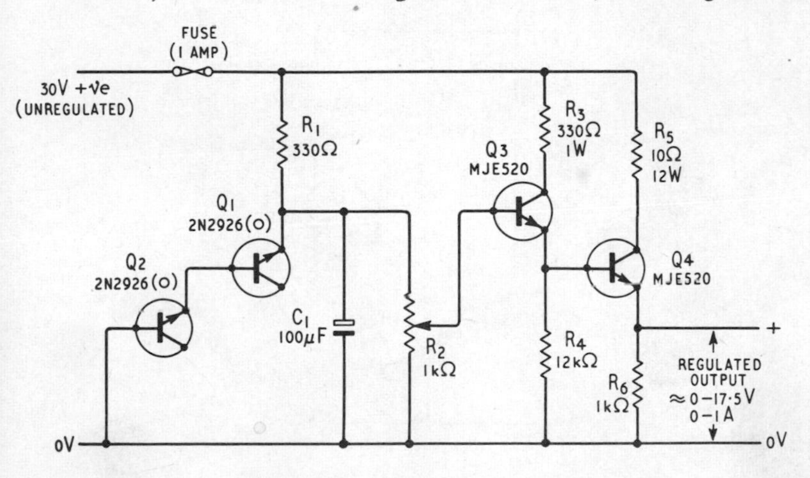

Fig. 1.18

Simple variable voltage regulator

supply must be derived from a transformer with fairly low copper losses, so that the full 30 V is available at maximum current load.

Current regulator circuits

The emitter and collector currents of a high gain transistor are inherently almost identical in amplitude, almost irrespective of the collector voltage, and it follows that the collector can thus be used as a constant-current source by simply setting the emitter current to the required value. This technique is of value in obtaining constant currents for charging DEAC batteries, for linearly charging capacitors in timer circuits, and for operating zener diodes as stable voltage reference sources. Fig. 1.19 shows the circuit of a practical current regulator working on this principle.

Here, *Q*1 is wired as a zener diode, and is operated at a current of about 9 mA via R_1. This zener voltage is fed to *Q*2 base, and so causes a regulated potential of about 7 V to appear at *Q*2 emitter; the emitter (and thus the collector) current of *Q*2 is thus dictated by this potential and by the combined resistance values of the emitter load resistors, R_2 and R_3, and can be varied over the approximate range 0.65–12.0 mA via R_2.

Thus, a constant current, of magnitude variable via R_2, is fed into any series load connected in the collector of *Q*2, and is independent

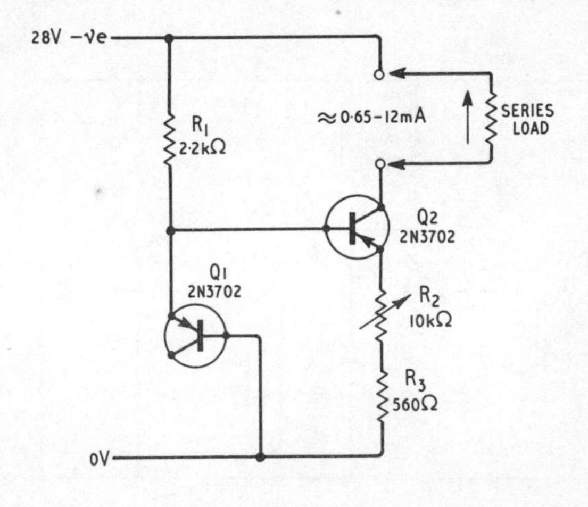

Fig. 1.19

Current regulator giving an output variable from approximately 0.65–12.0 mA

of the resistive value of the load providing it is not so large that the transistor is saturated.

In this circuit, the maximum available current is restricted to about 12 mA by the limited power dissipation capabilities of the 2N3702 transistor. Greater currents can be obtained, if required, by using a silicon power transistor in the *Q*2 position, and lowering the value of R_3 to limit the maximum current to the required value.

The circuit of Fig. 1.19 requires the use of a fixed 28 V supply. Fig. 1.20a shows the circuit of a constant current generator that can

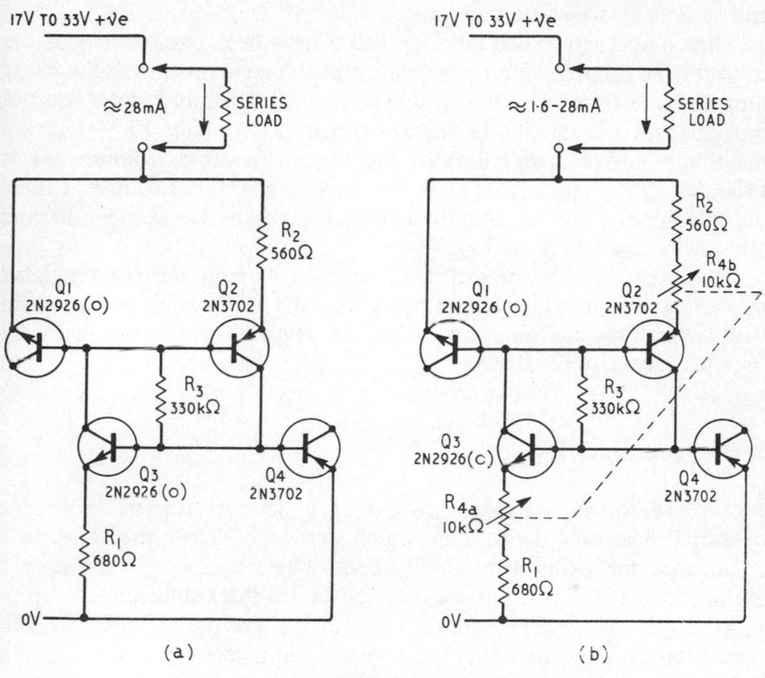

Fig. 1.20

(a) Constant current generator operating from a variable voltage supply. (b) Modification of Fig. 1.20a to give variable constant-current output

be operated from any supply in the range 17–33 V, and which draws a constant current of about 28 mA.

Here, *Q*1 is wired as a zener diode, and applies a fixed potential of about 9.5 V to *Q*2 base; *Q*2 has a fixed emitter load, R_2, of 560 Ω, so this transistor passes a constant collector current of about 17 mA.

This 17 mA current is fed to *Q*4, which is also wired as a zener diode and applies a fixed potential of about 7.5 V to *Q*3 base; *Q*3 has a fixed emitter load, R_1, of 680 Ω, so this transistor passes a constant collector current of about 11 mA, and this current is fed to zener diode *Q*1. Thus, both zener diodes are fed from constant current sources, and their operating potentials are well regulated. Consequently, the operating current of the entire circuit is fixed at about 28 mA, and is virtually independent of variations in supply line potential. R_3 prevents the transistors cutting off when the supply is first applied, and so acts as a sure-start resistor.

The component values of Fig. 1.20a have been chosen so that the circuit gives the maximum possible output current, within the working limits of the transistors used, i.e., with a 33 V supply and a shorted output load, the maximum voltage across *Q*3 is about 17 V, and the maximum power dissipations of the transistors are as follows: *Q*1 = 110 mW, *Q*2 = 290 mW, *Q*3 = 190 mW, and *Q*4 = 140 mW. Larger output currents can be obtained by using alternative semiconductors and lower values of R_1 and R_2.

The circuit can be modified to act as a variable current regulator by wiring a 2-gang 10 kΩ variable resistor in position as shown in Fig. 1.20b. This modification enables the regulated current to be varied over the range 1.6–28.0 mA.

Sine/square converter

Fig. 1.21a shows the basic circuit of a Schmitt trigger or voltage operated electronic switch, in which one or other of the transistors is on, and the other off, at all times. The values of R_1 and R_2 are chosen so that *Q*1 is normally off, and under this condition the top of R_4 (the *Q*2 base-bias resistor) is close to the +ve rail voltage, so *Q*2 is biased hard on and its collector is near ground volts. *Q*1 can be driven hard on by applying a positive signal to its base; under this condition the top of R_4 goes close to ground potential, and *Q*2 switches off, its collector going close to full +ve rail voltage. The values of R_3, R_6, and R_7 are chosen so that regenerative action occurs as the transistors change state. Thus, the circuit acts as an electronic switch which can be triggered from one state to the other by the application of a suitable input voltage.

This type of circuit can be used as a sine/square converter. When a large amplitude sine wave signal is applied to *Q*1 base, the +ve parts of the waveform cause *Q*1 to switch on, and the –ve parts cause *Q*1 to switch off again. Thus, a rectangular waveform appears at $Q1$ and $Q2$ collectors, and has a mark/space ratio of approximately 1:1, i.e., it resembles a square wave.

Fig. 1.21b shows the practical circuit of a sine/square converter.

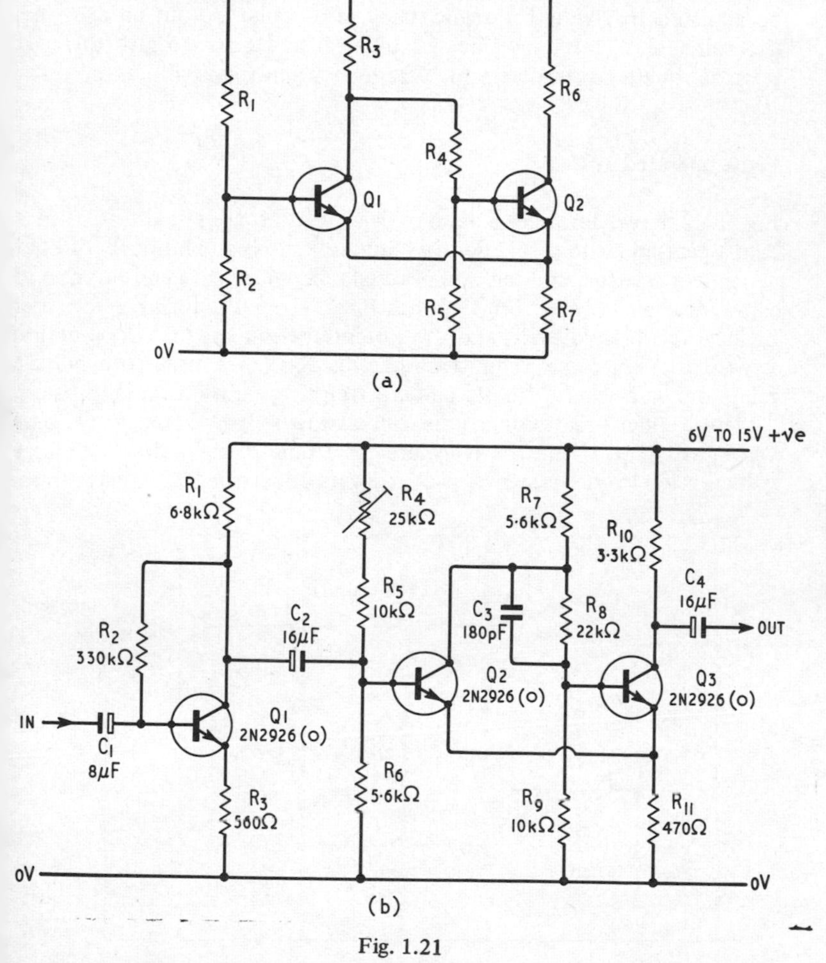

Fig. 1.21

(a) Basic Schmitt trigger. (b) Schmitt trigger used as a sine/square converter. Circuit has an input impedance of 40 kΩ, and requires a sine wave input greater than 100 $mV_{r.m.s.}$ for a square wave output

$Q1$ is wired as a simple common emitter pre-amplifier, and $Q2$ and $Q3$ are wired as a Schmitt trigger. The circuit can be used with any supply in the range 6–15 V, has an input impedance of about 40 kΩ, and requires a sine wave input of at least 100 $mV_{r.m.s.}$ to give a square wave output from $Q3$ collector via C_4. Good square waves are available over the frequency range of a few Hertz to over 100 kHz; R_4 should be adjusted to give a 1:1 mark/space ratio pulse output on a 'scope; the value of C_3 may be adjusted by trial and error to give the best possible square waves at very high frequencies, if required.

Light operated switch

Fig. 1.22 shows how the Schmitt trigger can be used as the basis of a light operated switch. L.D.R. is a cadmium sulphide photocell, or light dependent resistor, and has a high resistance under dark conditions and a low resistance under bright conditions. The l.d.r. forms a potential divider network with R_1, and the potential from the l.d.r.-R_1 junction is used to trigger the Schmitt circuit via R_2. $Q3$ is used to operate a relay, and is off when $Q1$ is off, and is driven to saturation when $Q1$ is on.

Thus, under bright conditions, only a low voltage is fed to $Q1$ base via R_2, so $Q1$, $Q3$, and the relay are off. Under dark conditions, a large voltage is fed to $Q1$ base via R_2 so $Q1$ triggers on, driving $Q3$ to saturation,

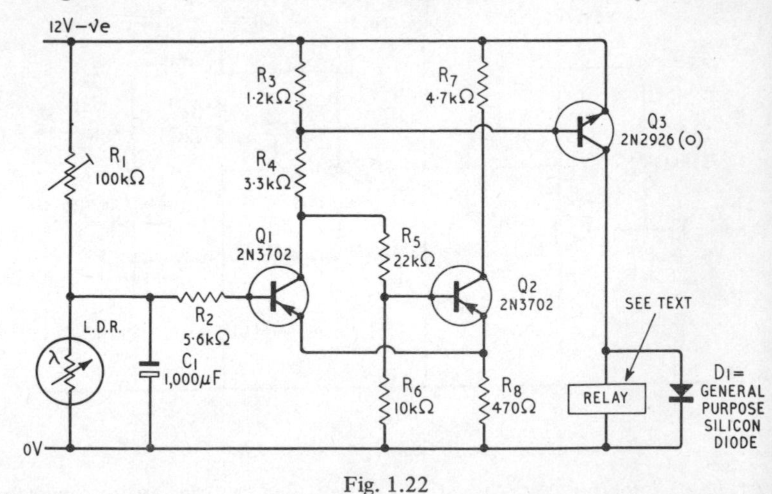

Fig. 1.22

Light operated switch, giving automatic operation of car parking or side lights. L.D.R. is any cadmium sulphide photocell with a face diameter greater than 0.25 in

and driving the relay sharply on. $D1$ is used to prevent any back e.m.f. from the relay coil damaging the circuit as the relay changes state.

The circuit is specifically designed to automatically operate car parking or side lights, via the relay contacts, and the precise trigger point can be adjusted via R_1. C_1 is included so that the circuit is operated by mean, rather than instantaneous, light levels, i.e., it is not effected by sudden changes in light levels, as might occur when driving under street lights, bridges, etc. The relay can be any 9–12 V type with a coil resistance greater than 270 Ω.

The circuit is intended for use in cars with 12 V +ve ground systems. It can be adapted for use in –ve ground systems by using 2N2926(0) transistors in place of the 2N3702 types, and vice versa, and by reversing the polarities of $D1$ and C_1.

Water operated switch

Fig. 1.23 shows how the circuit of Fig. 1.22 can be used with a +ve supply, and how it can be adapted as a water operated switch. In this case, the voltage that is fed to the base of $Q1$ via R_2 is taken from the emitter of emitter follower $Q4$. The base of $Q4$ is taken to ground via R_9, so that normally, with the metal probes isolated, there is

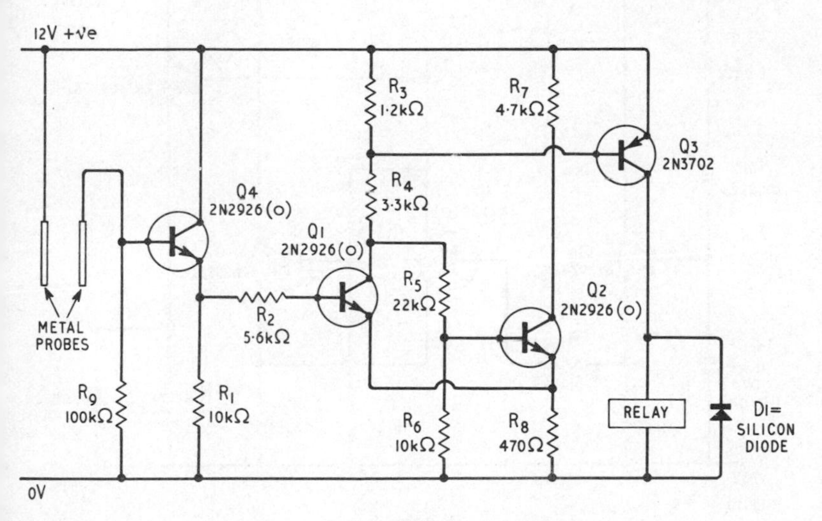

Fig. 1.23

Water operated switch

near-zero voltage on $Q1$ base, and the relay is off. If a resistance with a value less than about 330 kΩ is placed across the probes, however, potential divider action causes the emitter of $Q4$ to go sufficiently +ve to trigger $Q1$, and the relay then switches on.

Now, while it is true that distilled water has very good insulating properties, it is a fact that the impurities in normal tap water, or even in rain water in industrial areas, cause these liquids to have a fairly low resistance, so, in Fig. 1.23, the relay can be operated by placing the probes in normal water. The circuit has a number of applications in the home; it can, for example, be used to sound an alarm when bath water reaches a predetermined level, or to automatically wind in an outdoor washing line when it rains, etc.

Time switch

Fig. 1.24 shows how the circuit of Fig. 1.23 can be converted to a time switch, for use as a photo timer, etc. Here, R_9-R_{10} and C_1 are wired as a voltage divider network, so that, when the supply is connected,

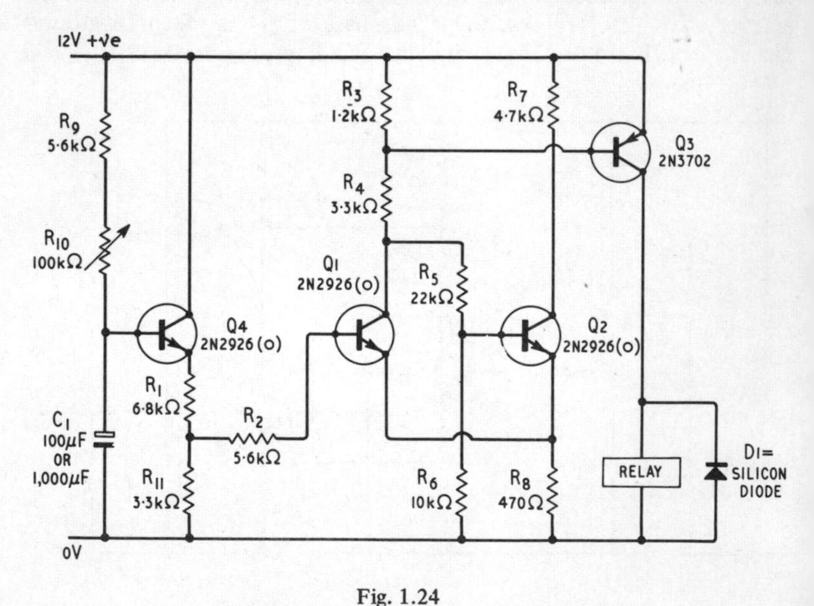

Fig. 1.24

Time switch

C_1 charges exponentially with a time constant that can be varied via R_{10} The rising exponential voltage is applied to $Q4$ base, and appears at the junction of R_1 and R_{11} at a reduced amplitude, and is then applied to the base of $Q1$ via R_2. The values of R_1 and R_{11} are chosen so that, with R_{10} set at maximum value, the Schmitt circuit triggers and the relay goes on after a delay of approximately 0.1 sec/μF value of C_1. This delay can be increased, if required, by increasing the value of R_1 or reducing the value of R_{11}.

With C_1 given a value of 100 μF, the delay can be varied from approximately 0.5 to 10 sec via R_{10}, and with a value of 1,000 μF it can be varied from about 5 sec to roughly 100 sec.

A.C. operated switch

A different type of electronic switch is shown in Fig. 1.25. Here, the relay operates when any a.c. input with an amplitude greater than about 100 $mV_{r.m.s.}$ is applied to $Q1$ base. The input impedance of the circuit

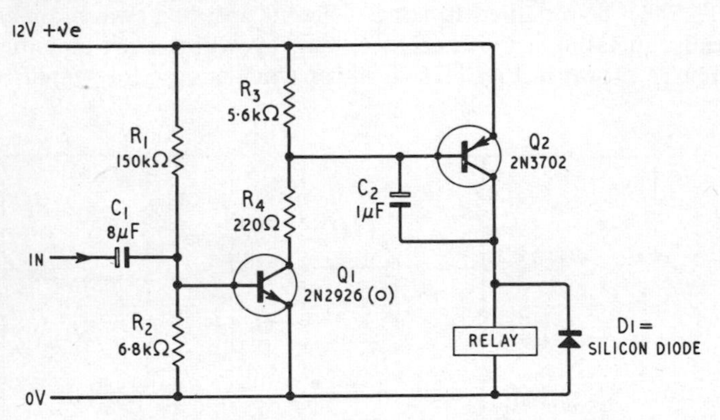

Fig. 1.25
A.C. operated switch, needing 100 $mV_{r.m.s.}$ to operate relay

is approximately 6 kΩ. The values of R_1 and R_2 are chosen so that a quiescent potential of about 0.5 V is applied to $Q1$ base, so, with no input signal connected, $Q1$, $Q2$ and the relay are off.

When an input signal with an amplitude greater than about 300 mV peak-to-peak (roughly 100 $mV_{r.m.s.}$ is applied to $Q1$ base, the +ve parts of the waveform drive $Q1$ on; as $Q1$ collector moves towards ground, the collector signal is partially smoothed by C_2, so a –ve going d.c. plus a.c. signal is applied to $Q2$ base, and that transistor is driven on. As $Q2$ is driven on, its collector moves towards +ve line potential, and the relay is driven on.

When the –ve going parts of the input waveform are applied to $Q1$ base, $Q1$ cuts off, but base current continues to flow in $Q2$ via C_2, so $Q2$ and the relay stay on. Thus, C_2 effectively converts the switching action of $Q1$ into a d.c. bias signal at $Q2$ base, and the circuit acts as an a.c. switch. R_4 prevents excessive base currents flowing in $Q2$.

As well as acting as a smoothing capacitor, C_2 also imparts a time delay to the on/off operation of the circuit; the duration of this delay depends on the values of both C_2 and the relay coil resistance. Long or short operating periods can be obtained by increasing or decreasing the value of C_2, to suit individual requirements. The relay can be any type with a coil resistance greater than about 180 Ω.

Sound operated switch

Fig. 1.25 can be modified to act as a sound operated switch, for automatically operating a tape recorder, etc., by wiring a pre-amplifier in position as shown in Fig. 1.26. In this particular case the amplifier of

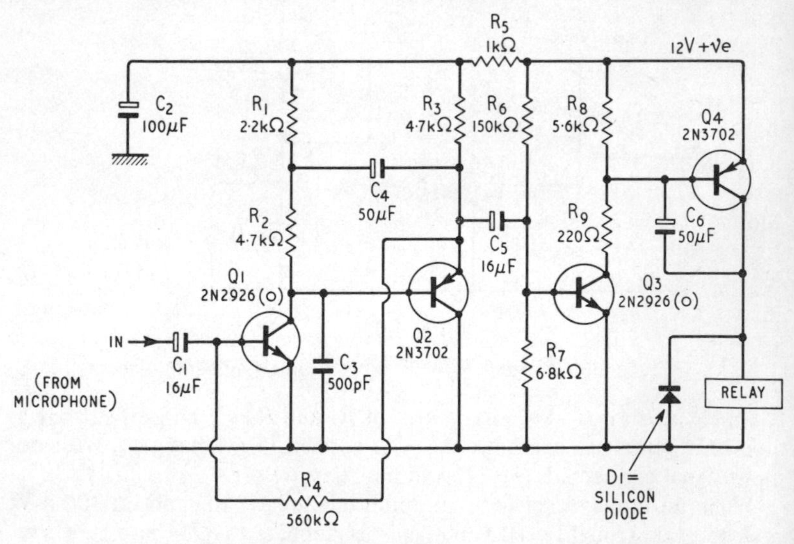

Fig. 1.26

Sound operated switch, needing 0.1 m$V_{r.m.s.}$ to operate relay

Fig. 1.7 has been used for this purpose, but other circuits are equally suitable. Decoupling network R_5-C_2 is wired between the two main sections of the unit, to prevent instability due to positive feedback. C_3 reduces the circuit's gain at high frequencies, and so prevents operation of the switch by stray signals picked up from the tape recorders bias oscillator.

C_6 is given a value of 50 μF in this application, and causes the switch to operate within about half a second of the input signal being applied, but delays switch-off by about 2.5 sec; these differences in switching times are mainly due to the differences in on and off operating voltages of the normal relay.

The circuit needs an input of about 0.1 $mV_{r.m.s.}$ to operate the relay, and is suitable for use with hand-held low to medium impedance microphones. Greater sensitivity can be obtained by using an additional pre-amplifier.

Tone operated switch

The circuit of Fig. 1.26 can be modified to act as a tone switch by incorporating a frequency selective network in the design, either at the input or in a –ve feedback loop. Fig. 1.27 shows a practical tone switch of this type, using a twin-T negative feedback element.

With the component values shown, the circuit is tuned to a centre frequency, f_o, of about 2.5 kHz, has an effective 'Q' of about 250, and needs an input of about 0.4 $mV_{r.m.s.}$ to operate the relay. When fed with a variable frequency input signal with an amplitude about 50% greater than that needed to operate the relay at the centre frequency, the unit exhibits a bandwidth of roughly ±2% of f_o, and is thus suitable for use in multi-channel remote control applications, etc.

The twin-T network (R_1-R_2-R_3-C_1-C_2-C_3) acts as a frequency-selective attenuator, with input applied to C_8 and output fed to Q1 base, and gives infinite attenuation at f_o, but low attenuation at all other frequencies. Thus, when connected in a negative feedback loop as shown, the amplifier gives a very high gain at f_o, but low gain at all other frequencies. For satisfactory operation (infinite attenuation at f_o), however, the twin-T components must be precisely matched in the following ratios:

$R_1 = R_2 = 2 \times R_3$, and $C_1 = C_2 = C_3/2$. In practice, the circuit gives good results if the twin-T components are matched to better than 5%.

The centre frequency, f_o, is approximately equal to $1/(6.3 \times R_1 \times C_1)$, so f_o can be reduced by increasing the resistor or capacitor values. R_1 and R_2 values can be varied over the range 4.7-22.0 kΩ; the non-standard R_3 values can be obtained by wiring two R_1 resistors in parallel.

The low frequency rejection characteristics of the circuit can be improved, if required, by reducing the values of C_4, C_6, and C_9, by trial

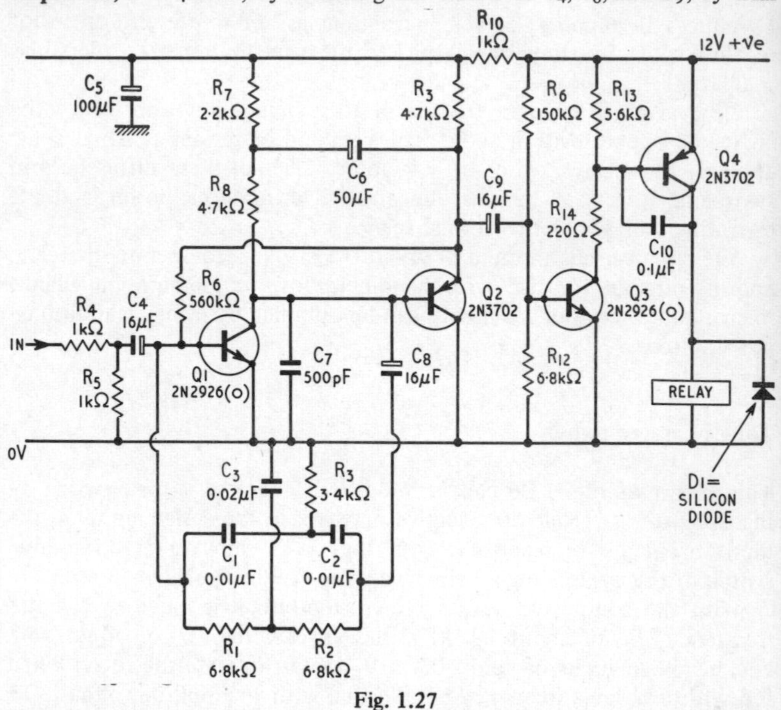

Fig. 1.27

2.5 kHz tone operated switch, needing 0.4 m$V_{r.m.s.}$ to operate relay. N.B. R_1, R_2, R_3, C_1, C_2 *and* C_3 *should be 5% or better*

and error. C_{10} has been given a value of 0.1 μF in this application, to make the unit suitable for use with pulsed tone signals, but this value can be varied to suit individual requirements. R_5 and C_7 are used to prevent positive current feedback at f_o, with consequent instability; their values may require adjustment at other frequencies, if stability is poor. The sensitivity of the circuit can be reduced, if required, by increasing the R_4 value.

Multivibrator circuits

Fig. 1.28 shows the circuit of a symmetrical 1 kHz astable multivibrator, or square wave generator. Outputs can be taken from either

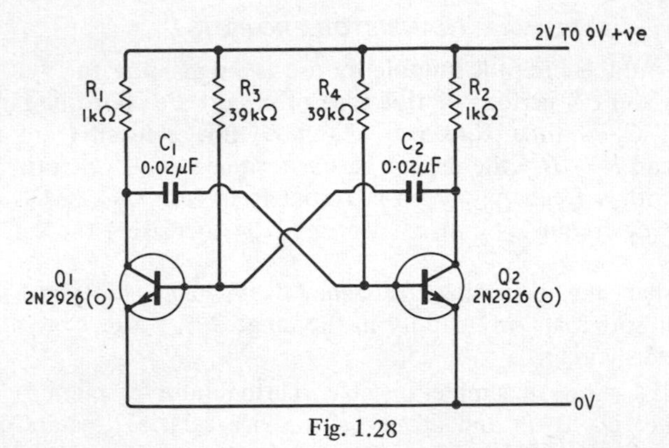

Fig. 1.28

Symmetrical 1 kHz multivibrator or square wave generator

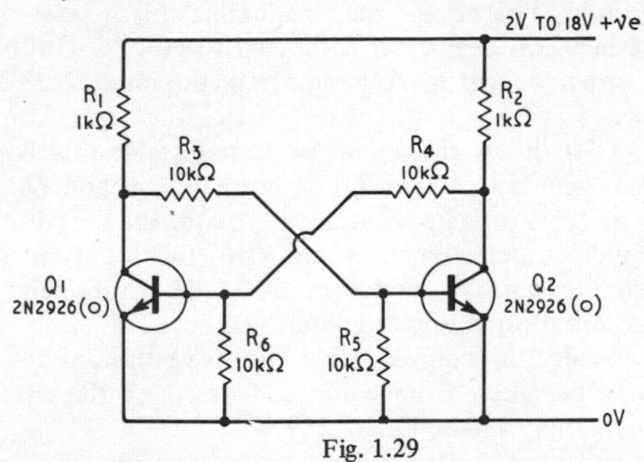

Fig. 1.29

Simple bistable multivibrator or memory unit

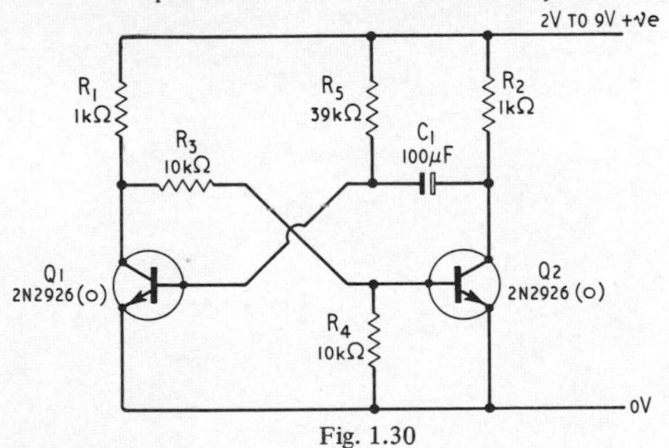

Fig. 1.30

Monostable multivibrator or one-shot pulse generator, giving 2.5 sec output pulse

collector, and the circuit is suitable for use as a signal injector.

The on and off periods of this type of circuit are controlled by the C_1-R_4 and C_2-R_3 time constants; if these time constants are equal ($C_1 = C_2$ and $R_3 = R_4$) the circuit acts as a square wave generator, and operates with a frequency of approximately $1/1.25 \times C_1 \times R_3$. Thus, the operating frequency can be decreased by increasing the values of C_1 and C_2.

Note that the operating frequency is virtually independent of supply rail potential. Any supply in the range 2–9 V can be used with this particular circuit.

Fig. 1.29 shows a simple bistable multivibrator or memory unit. Here, either $Q1$ is on and $Q2$ is off, or vice versa. The state of the circuit can be changed by momentarily shorting the base of the 'on' transistor to ground. The circuit then maintains this new state until the base of the new 'on' transistor is shorted to ground. Outputs can be taken from either collector. Any supply in the range 2–18 V may be used.

Finally, Fig. 1.30 shows the circuit of a monostable multivibrator, or one-shot pulse generator. Here, $Q1$ is normally on and $Q2$ is off; when the base of $Q1$ is briefly shorted to ground, the circuit changes state, but after a delay determined by the R_5-C_1 time constant returns automatically to the normal condition. With the component values shown, the pulse duration is approximately 2.5 sec. The circuit can be triggered electronically, if required, via a negative pulse applied to $Q1$ base. Outputs can be taken from either collector, and the circuit can be used with any supply in the range 2–9 V.

CHAPTER 2

15 FIELD-EFFECT TRANSISTOR PROJECTS

One of the most important new semiconductor devices to have been introduced in recent years is the field-effect transistor, or f.e.t. This device resembles a conventional transistor in a number of ways, but has the outstanding advantage of offering a very high input impedance at its 'gate'.

Two basic types of field-effect transistor are in use, and are known as the 'junction-gate f.e.t.' (JUGFET) and the 'insulated-gate f.e.t.' (IGFET) types. The IGFET type is, however, rather easily damaged if not carefully handled, so in this volume only the JUGFET type will be considered, and will be referred to simply as an 'f.e.t.'

An f.e.t., like an ordinary transistor, is a three-terminal device: the terminals are known as the 'source', the 'gate', and the 'drain', and correspond respectively to the emitter, base, and the collector of a normal transistor. 'N-channel' or 'p-channel' versions of the f.e.t. are available, just as normal transistors are available in either npn or pnp versions, and Fig. 2.1a shows the conventional symbols and supply polarities of both types of f.e.t. and of both types of ordinary transistor.

Like ordinary transistors, f.e.t.s can be used as amplifiers in any of three basic ways. Fig. 2.1b shows the three alternative modes of operation for npn transistors, (common emitter, common base, and common collector), and for n-channel f.e.t.s, (common source, common gate, and common drain).

F.E.T. characteristics

The most important characteristics of the f.e.t. are as follows:

(1) When an f.e.t. is connected to a supply with the polarity shown in Fig. 2.1a, (drain +ve for an n-channel f.e.t., –ve for

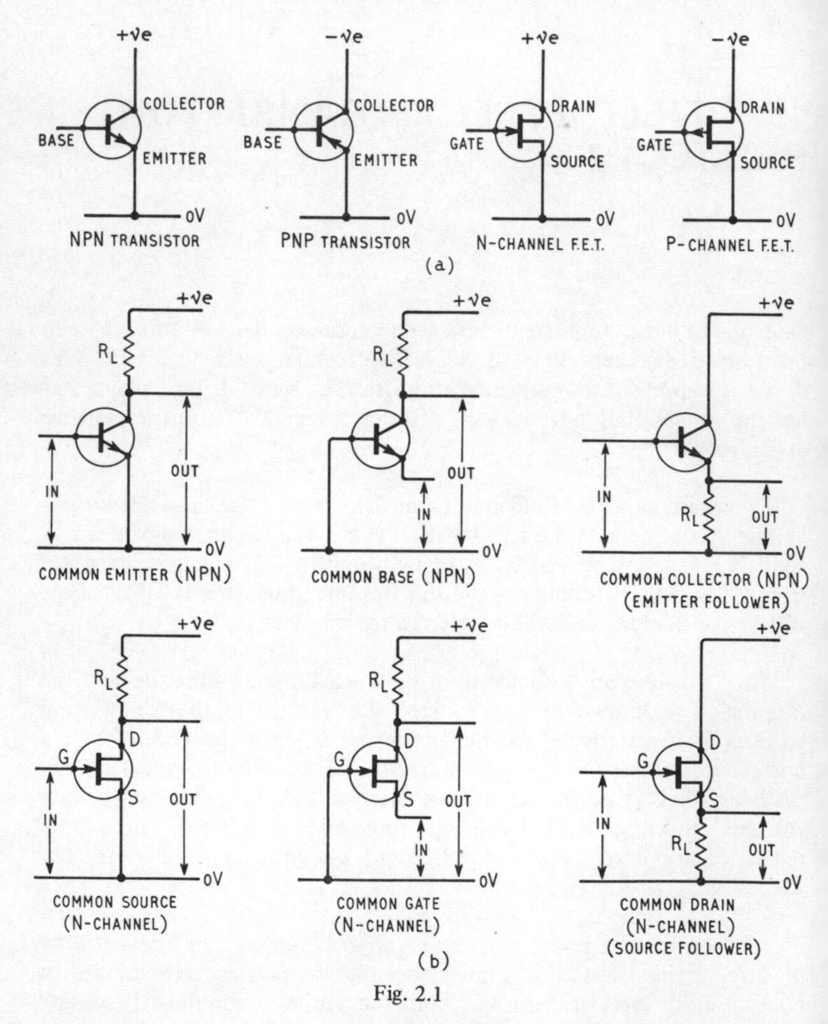

Fig. 2.1

(a) Transistor and f.e.t. symbols, with supply polarities. (b) The three basic transistor operating modes, and the f.e.t. equivalents

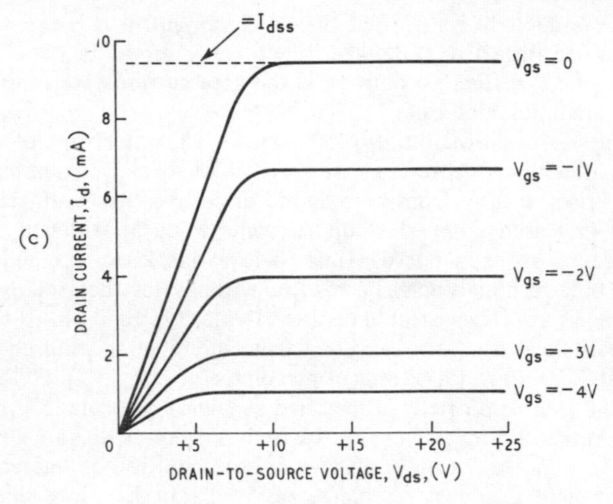

Fig. 2.1c
Typical transfer characteristics of n-channel f.e.t.

a p-channel f.e.t.), a drain current, I_d, flows in the device. The magnitude of I_d can be controlled via a gate-to-source bias voltage, V_{gs}.

(2) I_d is at a maximum when $V_{gs} = 0$, and is reduced (to bring the device into a linear operating region) by applying a reverse bias to the gate. Thus, a –ve gate voltage reduces I_d in an n-channel f.e.t., and a +ve bias has a similar effect in a p-channel device. The magnitude of V_{gs} needed to reduce I_d to zero is called the 'pinch-off' voltage, V_p, and typically has a value between 2 and 10 V. The magnitude of I_d when $V_{gs} = 0$ is denoted as I_{dss}, and typically lies between 2 and 20 mA.

(3) The gate-to-source junction of the f.e.t. has the characteristics of a silicon diode. When reverse biased (to bring the f.e.t. into a linear operating region), gate leakage currents, I_{gss}, are only a couple of nA (1 nA = 0.001 μA) at room temperatures; I_{gss} approximately doubles with every 10°C temperature rise, so only increases to a few μA at 125°C. Actual gate signal currents are only a fraction of an nA, and the input impedance to the gate is typically a thousand megohms at low frequencies; the gate junction is effectively shunted by a capacitance with a value of a few picofarads, so input impedances fall as frequency is increased. If the gate-to-source junction is forward biased,

it conducts like a normal silicon diode, and if it is excessively reverse biased it avalanches like a zener diode; in either case, the f.e.t. suffers no damage if the gate currents are limited to a few milliamperes.

(4) Fig. 2.1c shows the typical transfer characteristics of an n-channel f.e.t. Note that, for each value of V_{gs}, drain current I_d rises linearly from zero as the drain-to-source voltage, V_{ds}, is increased from zero up to some value at which a 'knee' occurs on each curve. Thus, below this knee, the drain-to-source terminals act as a resistor with a value dictated by V_{gs}, i.e., as a voltage-variable resistor. Typically, the drain-to-source resistance, R_{ds}, can be varied from a couple of hundred ohms (at $V_{gs} = 0$) to thousands of megohms (at $V_{gs} = V_p$).

(5) The gain of an f.e.t. is specified as transconductance, g_m, and signifies the rate of change of drain current with gate voltage, i.e., a g_m of 5 mA/V signifies that a variation of one volt on the gate produces a change of 5 mA in I_d. Note that the form I/V is the inverse of the ohms formula, so measurements specified in this way are usually expressed in 'mhos'. Usually, g_m is specified in f.e.t. data sheets either in terms of mmhos (milli-mhos) or μmhos (micro-mhos); thus, a g_m of 5 mA/V = 5 mmho = 5,000 μmho.

This completes the description of the general characteristics of the f.e.t. At this point, then, we can select a specific f.e.t. for experimental work, and then go on to consider a few practical circuits in which it can be used. The inexpensive 2N3819 n-channel f.e.t. has been selected for this purpose, and Fig. 2.2 and Table 2.1 show the general characteristics and lead connections of this particular device, which is encapsulated

Table 2.1

GENERAL CHARACTERISTICS OF THE 2N3819 F.E.T.

V_{DS}	=	+ 25 V (= max drain-to-source voltage).
V_{DG}	=	+ 25 V (= max drain-to-gate voltage).
V_{GS}	=	– 25 V (= max gate-to-source voltage).
V_p	=	– 8 V_{max} (= gate-to-source voltage needed to cut off I_d).
I_{dss}	=	2–20 mA (= drain-to-source current with $V_{gs} = 0$).
I_{gss}	=	– 2 nA_{max} (= gate cut-off (leakage) current at 25°C).
I_G	=	10 mA (= max gate current).
g_m	=	2.0–6.5 mmho (= small signal common source forward transconductance).
C_{iss}	=	8 pF_{max} (= common source short-circuit input capacitance).
P_T	=	200 mW_{max} (= power dissipation, in free air).
f_T	=	100 MHz (= gain-bandwidth product).

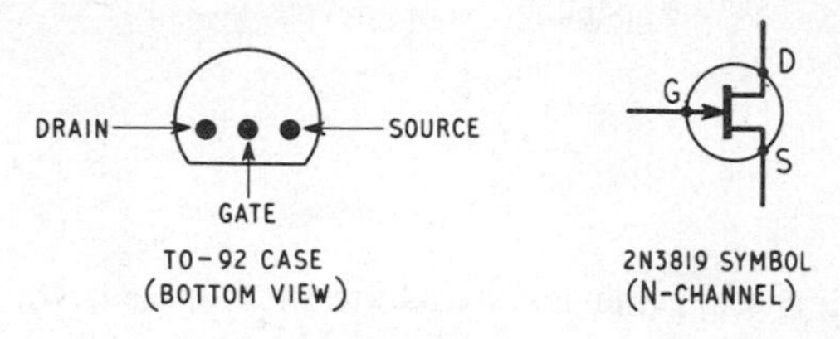

Fig. 2.2

Connections of the 2N3819 f.e.t.

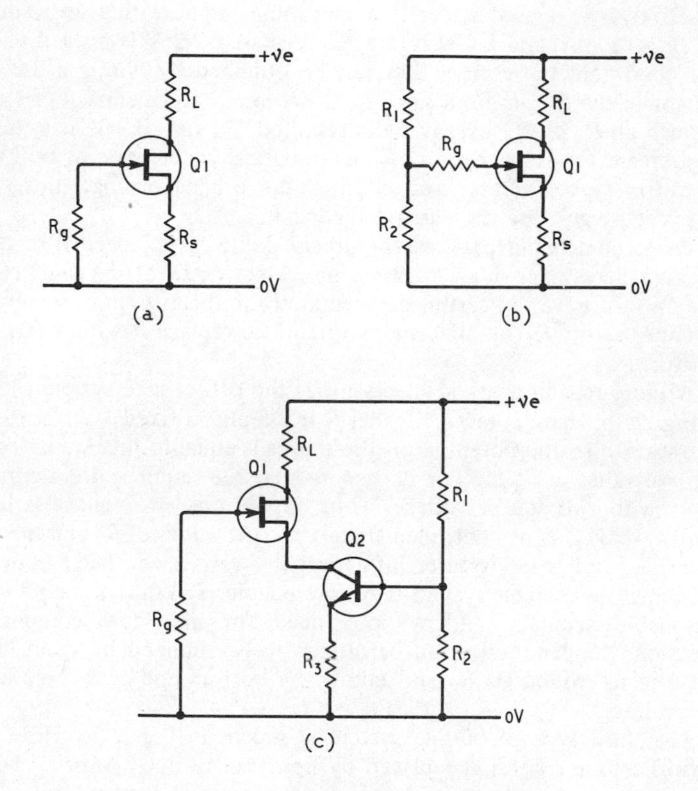

Fig. 2.3

(a) f.e.t. self-biasing system. (b) f.e.t. off-set gate biasing system. (c) Constant current f.e.t. biasing system

in an epoxy TO-92 package. The device is available from a number of manufacturers.

F.E.T. biasing

Three basic f.e.t. biasing systems are in use, each with its own particular advantages and disadvantages. The simplest of these is the self-biasing system shown in Fig. 2.3a. Here, the gate is tied to ground via R_g, and R_s is wired between the source and ground; any current flowing in R_s causes the source to go +ve relative to the gate, so the gate is effectively –ve biased under this condition. Suppose that we want to set I_d at 1 mA, and know that a V_{gs} bias of –2.2 V is needed to set this condition; the correct bias can be obtained by wiring a 2.2 kΩ resistor in the R_s position, since I_d flows in R_s, and a current of 1 mA through an R_s of 2.2 kΩ gives the required V_{gs} of –2.2 V. If I_d tends to decrease for some reason, V_{gs} automatically decreases as well and so causes I_d to increase and counter the original change; thus, the bias is self-regulating via negative feedback.

Unfortunately, in practice the precise value of V_{gs} needed to set a given I_d may vary widely between individual f.e.t.s of the same type: The only sure way of setting an accurate I_d in this system is, therefore, to either select R_s by trial and error, or to replace it with a variable resistor.

A more reliable method of biasing is the off-set gate system shown in Fig. 2.3b. Here, potential divider R_1-R_2 applies a fixed +ve bias to the gate via R_g, so the potential on the source is equal to this +ve bias plus the +ve value of V_{gs}; R_s is chosen so that the required drain current flows with this source voltage. Thus, if the +ve gate voltage is large relative to V_{gs}, I_d is controlled mainly by the values of R_s and the +ve gate bias, and is not greatly influenced by variations of V_{gs} between individual f.e.t.s. This system therefore enables I_d values to be set with reasonable accuracy and without need for individual component selection. Similar results can be alternatively obtained by connecting the gate to ground via R_g and taking the bottom end of R_s to a large –ve voltage.

The third type of biasing system is shown in Fig. 2.3c. Here, the normal source resistor is replaced by npn transistor $Q2$, which is wired as a constant current source and so determines the value of I_d. The value of this constant current is in turn determined by the voltage on $Q2$ base (set by potential divider R_1-R_2) and by the value of emitter resistor R_3; in some circuits, R_2 may be replaced by a zener diode or some other voltage reference device. Thus, in this system, I_d is independent of the f.e.t. characteristics, and very good stability is obtained, at the expense of increased circuit complexity and cost.

In the three biasing systems described, R_g can have any value up to about 10 MΩ, the maximum limit being imposed by the potential drop across this resistor caused by gate leakage currents, which may upset the biasing conditions.

Basic source follower circuits

The outstanding feature of the field-effect transistor is its inherently high input impedance, and full advantage can be taken of this characteristic when the device is used in the common drain or source follower mode (the f.e.t. equivalent of the emitter follower mode). Fig. 2.4a shows a simple circuit of this type.

Here, a self-biasing system is used, and the drain current can be varied via R_1. The circuit can be used with any supply in the range

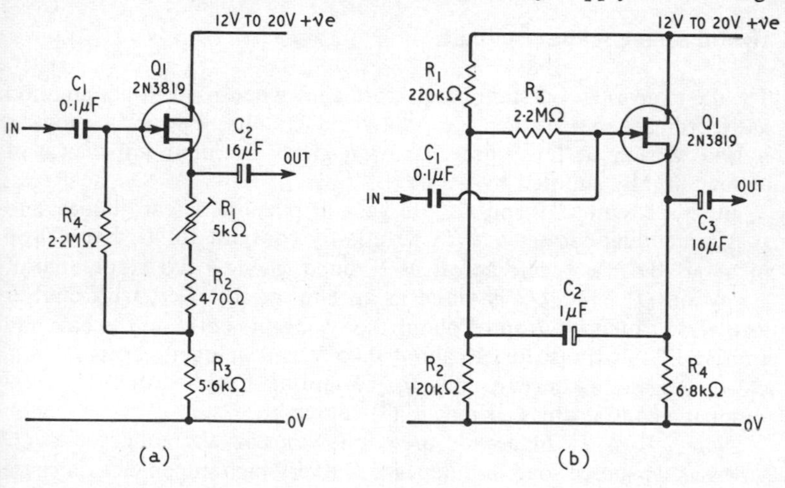

Fig. 2.4

(a) Simple source follower, giving:

A_v = *0.95*

Z_{in} = *10 MΩ shunted by 10 pF*

(b) Simple source follower, giving:

A_v = *0.95*

Z_{in} = *44 MΩ shunted by 10 pF*

12–20 V, and R_1 should be adjusted so that the quiescent potential across R_3 is 5.6 V, giving a drain current of 1 mA. The circuit gives a voltage gain of 0.95 between input and output.

Due to the potential divider action of the R_1-R_2 to R_3 chain, a degree of bootstrapping is applied to R_4, and its effective value is increased by about 5 times. The actual input impedance of the circuit is about 10 MΩ shunted by 10 pF, i.e., it is 10 MΩ at very low frequencies, falling to 1 MΩ at about 16 kHz, and to about 100 kΩ at 160 kHz.

Fig. 2.4b shows an alternative version of the simple source follower circuit. In this case, gate off-set biasing is used, so individual component adjustment is not required. Voltage gain is approximately 0.95. C_2 is a bootstrap capacitor, and increases the effective value of gate resistor R_3 by about 20 times. C_2 can be omitted from the design, if preferred.

With C_2 removed from the circuit, the input impedance of the design is about 2.2 MΩ shunted by 10 pF. With C_2 in place, the input impedance is raised to about 44 MΩ shunted by 10 pF. Alternative impedance values can be obtained by changing the R_3 value, up to a maximum of 10 MΩ.

Hybrid source follower circuits

The f.e.t. gives an outstanding performance when used in conjunction with ordinary transistors, i.e., in hybrid circuits, Fig. 2.5a shows a hybrid version of the source follower, giving an input impedance of about 500 MΩ shunted by 10 pF.

In this circuit, *D*1 and *D*2 are general purpose silicon diodes, and pass a standing current via R_5, so a fairly constant forward volt drop of about 0.65 V occurs across each diode, giving a fixed potential of 1.3 V on *Q*2 base. *Q*2 is wired as an emitter follower, with emitter load R_4; a potential drop of about 0.65 V occurs between the base and emitter of this transistor, so about 0.65 V is developed across R_4, and *Q*2 thus passes a constant collector current of roughly 1 mA. Thus, *Q*2 supplies a constant bias current to *Q*1 source.

Now, *Q*1 is wired as a source follower, and the collector of *Q*2 serves as its source load and appears as a very high impedance. Because of the very high effective value of this source load, the f.e.t. gives a voltage gain of about 0.99. C_2 passes a bootstrap signal from *Q*1 source to R_3, and because of the high voltage gain of the circuit this bootstrap signal increases the effective value of R_3 by about 100 times, i.e., to 1,000 MΩ. Thus, the actual input impedance of the circuit is equal to this value shunted by the f.e.t.s gate impedance, and works out at about 500 MΩ shunted by. 10 pF.

If the high effective value of source load (and thus the high input impedance) of this circuit is to be maintained, the output must either be taken to external circuits via an additional emitter follower, or taken only to fairly high impedance loads.

Fig. 2.5b shows how a pnp emitter follower, $Q3$, can be added to

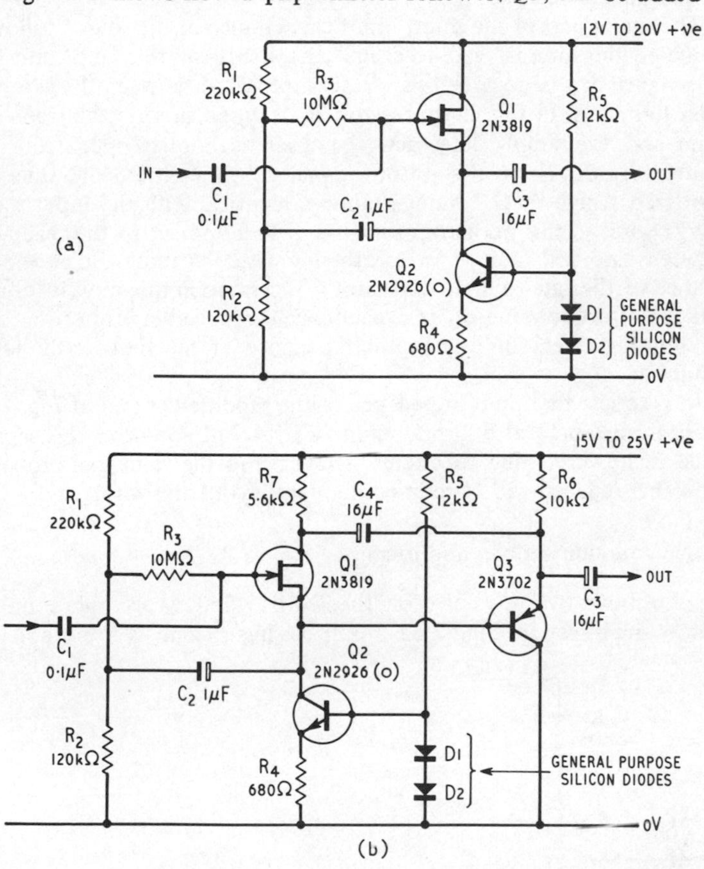

Fig. 2.5

(a) Hybrid source follower, giving:

A_v = *0.99*

Z_{in} = *500 MΩ // 10 pF*

(b) Modified source follower, giving:

A_v = *0.99*

Z_{in} = *500 MΩ // 4.7 pF*

the above circuit, enabling an output to be taken directly to a low impedance external load via C_3. In addition, this particular circuit incorporates modifications that reduce the effective shunt capacitance of the input impedance, and so enables an improved high frequency performance to be obtained.

The major part of the shunt input capacitance of the source follower is due to the internal gate-to-drain capacitance of the f.e.t., and this capacitance can be regarded as a reactance wired between the gate and drain terminals. In Fig. 2.5b, resistor R_7 is wired in series between the drain and +ve supply line, and the drain is bootstrapped from $Q3$ emitter via C_4. Now, the output signal at $Q3$ emitter, and thus the bootstrap signal on $Q1$ drain, is almost identical with the input signal on $Q1$ gate, so this bootstrap signal is in fact applied to the reactance between gate and drain, and greatly increases its value. Since the reactance of the gate-to-drain capacitor is increased in this way, it follows that the effective value of its capacitance is reduced in proportion, and its shunting effect on the input impedance of the f.e.t. is therefore minimised.

In practice, the input impedance of the modified circuit of Fig. 2.5b has been measured at 500 MΩ shunted by 4.7 pF. Some of this capacitance is, however, due to circuit 'strays', and the value can probably be further reduced with care in component layout and wiring.

Simple common source amplifiers

Fig. 2.6 shows two ways of using the 2N3819 f.e.t. as a simple common source amplifier. In Fig. 2.6a a self-biasing system is used, and the

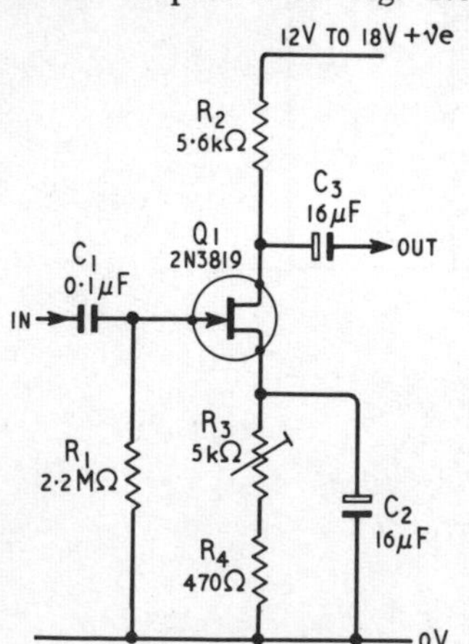

Fig. 2.6a

Simple common source amplifier with self-biasing, giving:

A_V = *21 dB*
Z_{in} = *2.2 MΩ // 50 pF*
f_R = *15 Hz–250 kHz ± 3 dB*

circuit can be used with any supply in the range 12–18 V R_3 should be adjusted so that 5.6 V is developed across R_2.

Fig. 2.6a shows the off-set gate biasing version of the same circuit. In this case, the supply range is limited to between 18 and 20 V; I_d is

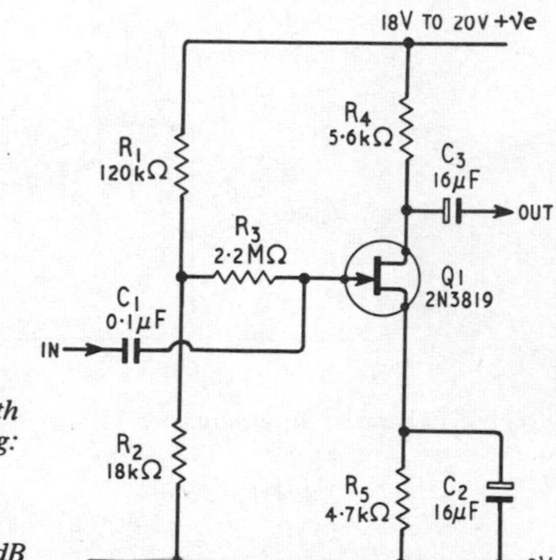

Fig. 2.6b
Simple common source amplifier with off-set gate biasing, giving:

A_V = *21 dB*
Z_{in} = *2.2 MΩ // 50 pF*
f_R = *15 Hz–250 kHz ± 3 dB*

approximately 1 mA. In both of these circuits, the source is decoupled to ground via C_2 at signal frequencies.

These two circuits give a similar small signal performance, although this is subject to some variation between individual f.e.t.s. On average, the voltage gain of both circuits works out at 21 dB (= approximately 12 times), and the frequency response is within 3 dB from 15 Hz to 250 kHz. Input impedance is about 2.2 MΩ shunted by 50 pF. This comparatively high value of shunt capacitance is due to Miller feedback from drain to gate, which effectively increases the value of the f.e.t.s gate-to-drain capacitance in proportion to the voltage gain of the amplifier, i.e., by 12 times in this particular case.

Hybrid common source amplifier

Fig. 2.7 shows the hybrid version of the simple common source amplifier, using npn transistor $Q2$ as a constant current bias supply for the f.e.t; the source of $Q1$ is decoupled to ground via C_2.

This version of the amplifier has excellent bias stability, and is

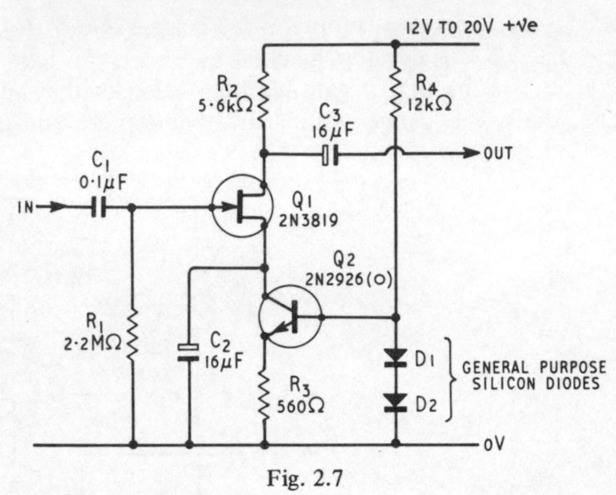

Fig. 2.7

Hybrid common source amplifier, giving:

A_v = *21 dB*
Z_{in} = *2.2 MΩ // 50 pF*
f_R = *15 Hz–250 kHz ± 3 dB*

suitable for use with any supply in the range 12–20 V. Its small signal performance is identical with that of the amplifier shown in Fig. 2.6a.

A compound amplifier

Each of the three common source amplifiers described above has a fairly large value (50 pF) of shunt input capacitance, and thus has a moderately low value of input impedance at high frequencies (approximately 32 kΩ at a frequency of 100 kHz). Fig. 2.8 shows an alternative version of the common source amplifier, in which the input shunt capacitance is substantially reduced (to about 12 pF), thus giving an increased input impedance at high frequencies (about 120 kΩ at 100 kHz). This circuit uses a variety of transistor operating modes (common source, common base, and common collector), and is therefore known as a 'compound amplifier'.

The cause of the large input shunt capacitance of the conventional common source amplifier is the Miller effect, which increases the f.e.t.s gate-to-drain capacitance in proportion to the voltage gain between gate and drain. In Fig. 2.8, the f.e.t. (*Q*1) is wired as a normal common source amplifier, with constant current biasing provided via *Q*2, but in this case the drain of the f.e.t. is connected directly to the emitter of a third transistor, $Q3$, which is wired as a common base amplifier. Now, as far as the f.e.t. is concerned, the emitter of $Q3$ appears as a very low impedance drain load, which effectively couples the drain to

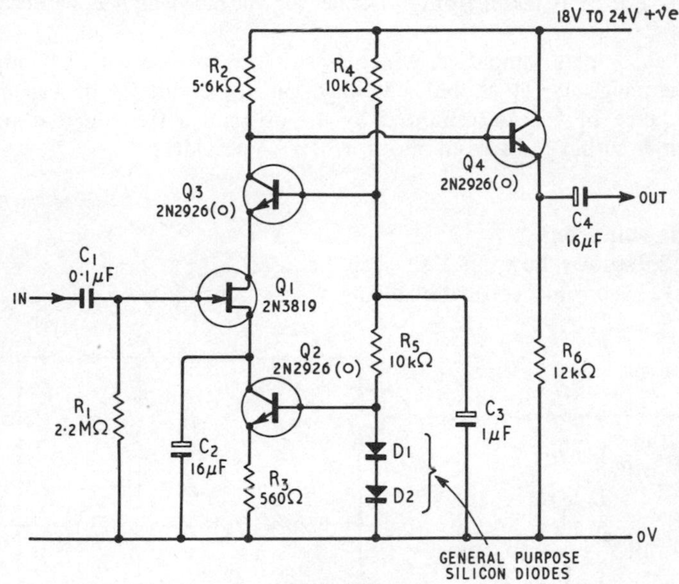

Fig. 2.8

Compound amplifier, giving:

A_V = *21 dB*
Z_{in} = *2.2 MΩ // 12 pF*
f_R = *15 Hz–1.5 MHz ± 3 dB*

ground at signal frequencies via the forward biased emitter-base junction of $Q3$ and via C_3; consequently, only negligible voltage amplification occurs between the gate and drain of $Q1$, and there is very little increase in the f.e.t.s gate-to-drain capacitance as a result of the Miller effect. $Q1$ therefore exhibits a fairly low value of input shunt capacitance.

Although only negligible voltage amplification takes place between the gate and drain, current amplification takes place in the normal way, and the drain signal currents are fed directly into the emitter of the common base amplifier, $Q3$, which uses collector load R_2. $Q3$ has near-unity current gain between emitter and collector, so the signal currents flowing in R_2 and in the drain of $Q1$ are virtually identical;

consequently, reasonable voltage gain (about 21 dB) occurs between $Q1$ gate and $Q3$ collector. To prevent external circuits with fairly large values of shunt input capacitance reducing the effective value of R_2 (and thus the amplifier's voltage gain) at high frequencies, the output of the circuit is taken from $Q3$ collector via $Q4$, which is wired as an emitter follower.

The complete amplifier, which is suitable for use with any supply in the range 18–24 V, has a voltage gain of about 21 dB, an input impedance of 2.2 MΩ shunted by 12 pF, and a frequency response which is within 3 dB from about 15 Hz to 1.5 MHz.

F.E.T. voltmeters

Fig. 2.9 shows how an f.e.t. can be used as the basis of a simple 3-range electronic voltmeter, giving a basic sensitivity of 22.2 MΩ per

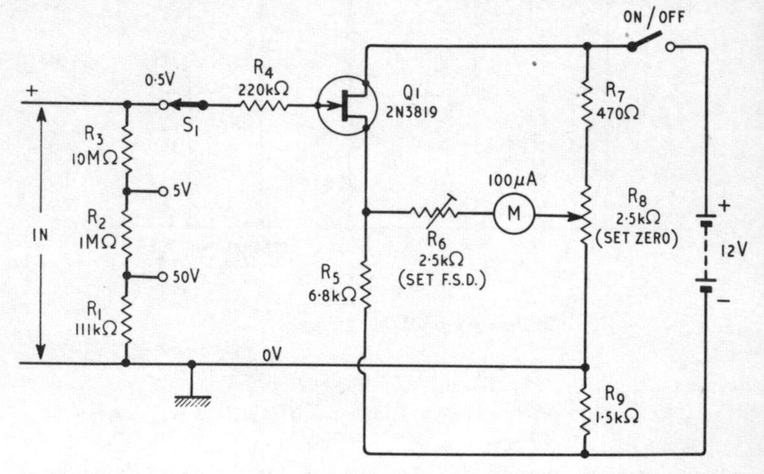

Fig. 2.9
Simple 3-range f.e.t. voltmeter

volt. Maximum full scale voltage sensitivity is 0.5 V, and input resistance is constant at 11.1 MΩ on all ranges.

R_7-R_8 and R_9 form a potential divider across the 12 V battery, and cause 4 V to appear across R_9; the top end of R_9 is connected to the ground of the circuit, which can be regarded as a zero volts line, so the bottom end of R_9 is at a potential of 4 V –ve, and the top of R_7 is at 8 V +ve. $Q1$ is wired as a source follower, with its gate taken to ground via the R_1 to R_4 network, but the source of $Q1$ is connected to the 4 V –ve line via source load R_5, so the f.e.t. is effectively given off-set gate biasing, and its drain current is automatically set at about 1 mA.

R_7-R_8 and $Q1$-R_5 act as a bridge circuit, and R_8 is adjusted so that, in the absence of an input voltage at $Q1$ gate, the voltage on $Q1$ source is equal to that on R_8 slider, so the bridge is balanced and zero current flows in the meter. Any potential applied to $Q1$ gate then causes the bridge to go out of balance by an amount proportional to the input voltage, which can then be read directly on the meter. R_1-R_3 form a simple range multiplier network, giving full scale deflection ranges of 0.5 V, 5 V, and 50 V. Alternative networks can be used if preferred, but close tolerance components should be used if good accuracy is required. R_4 acts as a safety resistor, and prevents damage to $Q1$ gate in the event of excessive input voltages being connected.

In use, R_8 is first adjusted so that the meter reads zero in the absence of an input voltage. An accurately known potential of 0.5 V is then connected to the gate, and R_6 is adjusted to give a full scale deflection on the meter. These adjustments are then repeated until consistent zero and full scale deflection readings are obtained, and the unit is then ready for use.

In practice, this unit is rather prone to drift with changes in temperature and supply voltage, so frequent re-adjustment of the zero control is required; drift can be considerably reduced by using a zener stabilised 12 V supply.

A low drift version of the f.e.t. voltmeter is shown in Fig. 2.10. Here, $Q1$ and $Q2$ are wired as a differential amplifier, so any drift occuring on one side of the circuit is automatically countered by a similar drift on the other side, and very good stability is obtained. The circuit works on the bridge principle; $Q1$-R_5 form one arm of the bridge, $Q2$-R_6 the other.

It is important to note that $Q1$ and $Q2$ are selected f.e.t.s. in this circuit, and must have their I_{dss} values matched within 10%. The circuit can be used with any supply in the range 12–18 V, and the setting up procedure is similar to that described for the Fig. 2.9 circuit.

Very low frequency astable multivibrator

Fig. 2.11 shows the circuit of a very low frequency f.e.t. astable or free-running multivibrator. The on and off periods of the circuit are controlled by the time constants C_1-R_4 and C_2-R_3; because of the ultra-

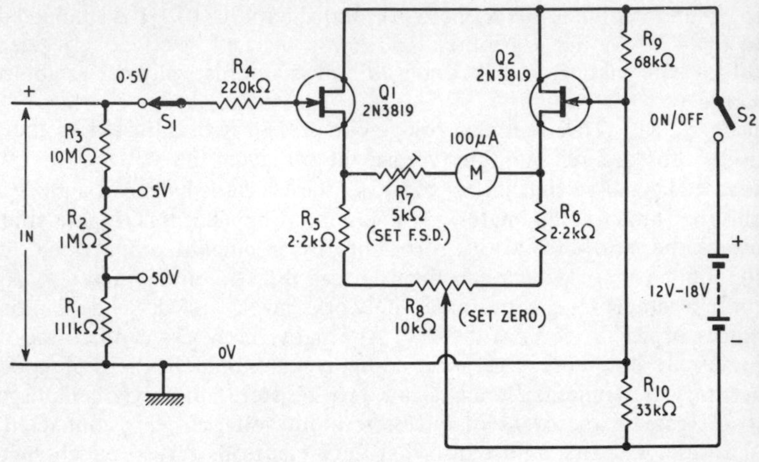

Fig. 2.10

Low-drift 3-range f.e.t. voltmeter. Q1 and 2 must have I_{dss} values matched within 10%

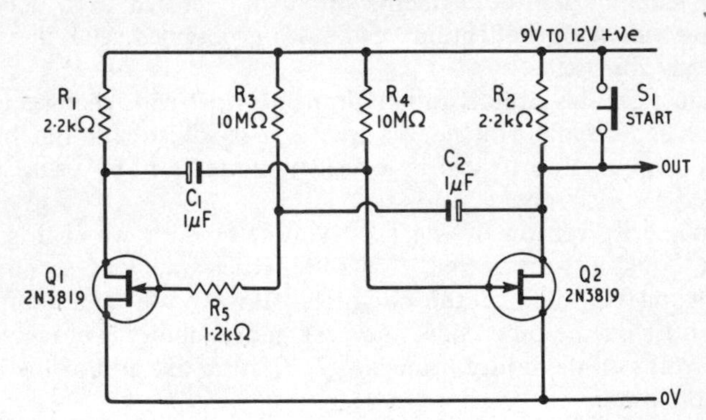

Fig. 2.11

V.L.F. astable multi, with cycling rate of one in 20 sec

high input impedances of the f.e.t.s, the *'R'* parts of these time constants car. be made very large, so that very long cycling periods can be obtained using fairly low values of *'C'*. With the component values shown, the prototype circuit cycled at a rate of once every 20 sec, i.e., at a frequency of 0.05 Hz. A second version of the circuit, using 40 μF components in the C_1 and C_2 positions, cycled at a rate of once every 6 min. C_1 and C_2 must be low-leakage capacitors, such as Mylar, Tantalum, etc.

The operating principle of the circuit is similar to that of the normal transistor astable multi, except that in the f.e.t. case it is necessary to apply a charging voltage to C_2, by closing the 'start' button for about 1 sec, to initiate circuit operation. R_5 ensures that excessive gate currents do not flow in $Q1$ when the 'start' button is operated.

The on and off periods of the circuit can be made variable, if required, by replacing both R_3 and R_4 with a 10 MΩ variable resistor in series with a 1 MΩ fixed resistor. The values of C_1 and C_2 can be increased or decreased to suit individual requirements.

Timer circuits

Field effect transistors are suitable for use in a variety of electronic timer circuits, and Fig. 2.12 shows one such example. With C_1 given a value of 1 μF, the prototype circuit gives a timing period of 40 sec, and with a value of 100 μF it gives a period of 35 min.

In this circuit, $Q1$ is wired as a source follower, and has its gate taken to the junction of time constant network R_1-C_1. When the supply is first connected, C_1 is discharged, so $Q1$ gate is at ground potential, and the source is a volt or two higher; the base of pnp transistor $Q2$ is connected to $Q1$ source via R_3, so $Q1$ is driven on under this condition, and a 12 V output appears across R_5.

As soon as the supply is connected, C_1 starts to charge via R_1, so the voltages on $Q1$ gate and source rise exponentially towards the 12 V

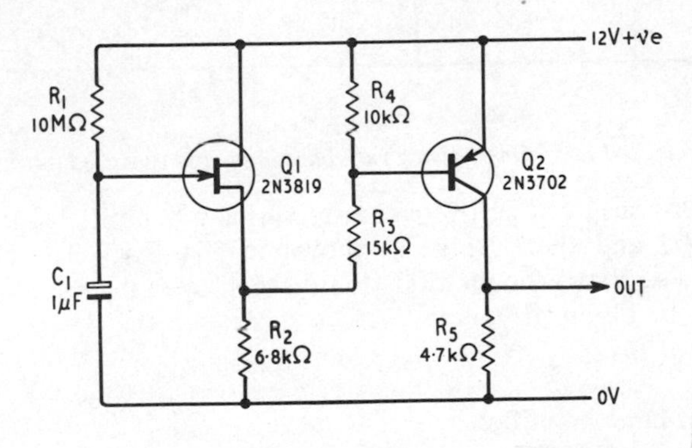

Fig. 2.12

Simple f.e.t. timer, giving a period of 40 sec when $C_1 = 1$ μF, and 35 min when $C_1 = 100$ μF

supply line; eventually, when $Q1$ source rises to about 10.5 V, the forward bias of $Q2$ falls to zero and $Q2$ switches off; zero volts output appears across R_5 under this condition.

When the supply is removed from the circuit, C_1 discharges rapidly via R_2 and the forward biased internal gate-to-drain junction of $Q1$, and the circuit is then ready to carry out a second timing operation as soon as the supply is re-connected.

The circuit can be made to give a variable timing period by replacing R_1 with a 10 MΩ variable resistor and 1 MΩ fixed resistor in series. C_1 should be a Mylar or similar low-leakage type of capacitor.

A minor disadvantage of the circuit of Fig. 2.12 is that $Q2$ switches off rather slowly, so a sharp switching action is not obtained at the

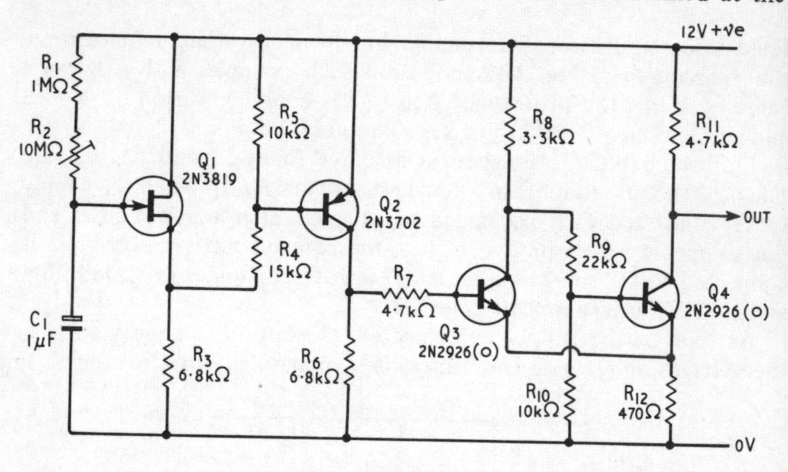

Fig. 2.13
Modified f.e.t. timer, giving variable timing and switched output

output. This snag can be overcome by wiring a Schmitt trigger circuit between $Q2$ and the output, as shown in Fig. 2.13. The circuit of Fig. 2.13 also shows the modifications needed to enable variable timing periods to be obtained.

Constant-volume amplifier

When operated with a low drain voltage, the drain-to-source path of the n-channel f.e.t. exhibits the characteristics of a simple resistor, the value of which can be varied via a negative bias applied to the gate;

this resistance is low when zero gate bias is applied, and very large when a substantial negative gate bias is applied.

This characteristic of the f.e.t. makes it suitable for use in variable voltage-operated attenuator networks, and Fig. 2.14 shows how such

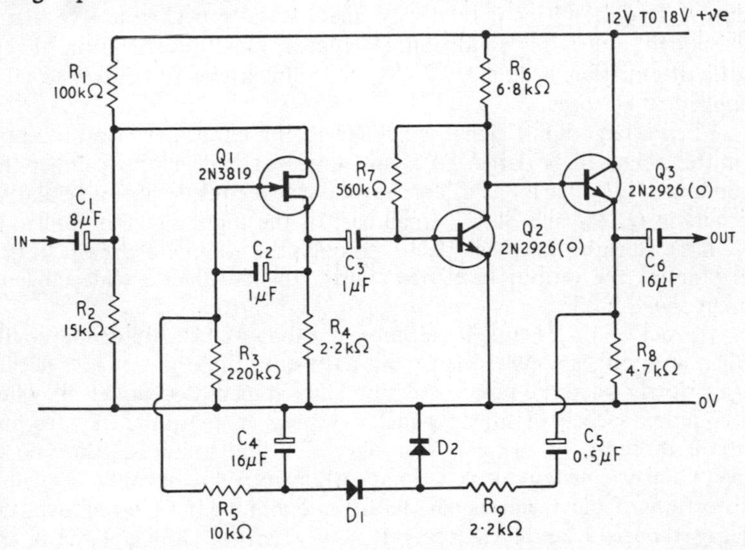

Fig. 2.14

Constant-volume amplifier, giving 7.5 dB change in output for 40 dB change in input. N.B. D1 *and* D2 *are general purpose germanium diodes*

a network can be used as the basis of a constant-volume amplifier. With 300 $mV_{r.m.s.}$ applied to the input of the unit, an output of 0.72 V is available on the prototype, and when the input is reduced to 3 mV the output falls to 0.3 V, i.e., a 40 dB change in the input signal level produces a change of only 7.5 dB at the output. Providing that inputs are kept to less than 500 mV, the circuit gives very little distortion.

In this circuit, *Q*1 and R_4 are wired as a voltage operated attenuator, with input applied to *Q*1 drain via C_1, and output taken from *Q*1 source via C_3; a small positive voltage is applied to the drain via R_1 and R_2. The output from *Q*1 source is fed to the *Q*2-*Q*3 common emitter/emitter follower amplifier, and the output of *Q*3 emitter is fed, via C_5 and R_9, to the *D*1-*D*2-C_4 rectifier/smoothing network, so that a –ve potential is developed across C_4, and is proportional to the signal amplitude at *Q*3 emitter. This –ve potential is applied to *Q*1 gate, and so controls the attenuation of $Q1$-R_4. C_2 ensures that the gate-to-source bias is not modulated by the output of the attenuator, and thus keeps distortion low.

When a very small signal is applied to the input of the circuit, the output at $Q3$ emitter is relatively small, so only negligible –ve bias is developed; under this condition, $Q1$ appears as a low resistance, so very little attenuation occurs in $Q1$-R_4, and almost the full input signal is applied to $Q2$ base.

When a large input signal is applied to the circuit, the output at $Q3$ emitter tends to be large, so a large –ve bias is developed; under this condition, $Q1$ appears as a large resistance, so considerable attenuation occurs in $Q1$-R_4, and only a small part of the input signal is applied to $Q2$ base. Negative feedback occurs through the complete circuit, so that in practice the output level stays fairly constant over a wide range of input signal levels.

The action of $D1$ and $D2$ ensures that the –ve bias builds up rapidly when an input is applied via C_5, but C_4 ensures that the –ve bias decays again slowly when the input is removed or reduced. Consequently, when a complex speech or music signal is applied to the unit, the –ve bias circuit responds to the peaks of the signal and so adjusts the gain to give a fairly constant peak volume, while introducing only negligible distortion to the mean signal. With the component values shown, the decay time is a couple of seconds, and when the unit is wired in the a.f. stages of a radio receiver it makes it possible to tune through a complete waveband without need to adjust the volume control, both strong and weak stations appearing at equal volume.

For normal listening on a single station, the value of C_4 should be increased to 100 μF, to increase the decay time of the –ve line. The

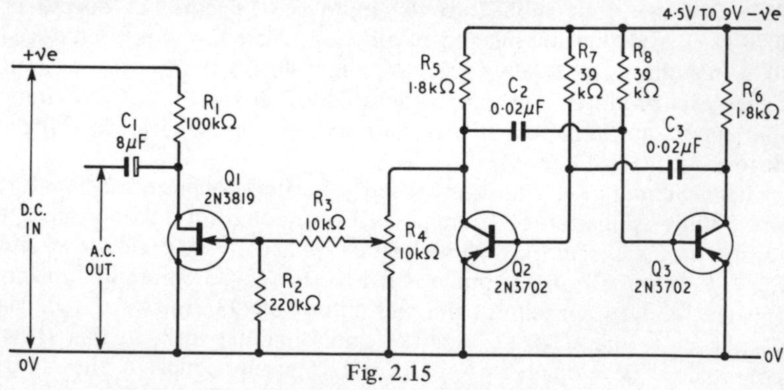

Fig. 2.15

f.e.t. chopper or d.c. to a.c. converter

circuit then eliminates the 'fade' that occurs on distant stations, but does not introduce excessive automatic volume adjustment during brief pauses in normal speech. This characteristic is also useful in tape recorder and intercom circuits, etc.

F.E.T. chopper

Finally, Fig. 2.15 shows how the low voltage resistor-like characteristics of the f.e.t. can be used in a chopper application, to convert a d.c. input voltage into a square wave 'chopped' output with an amplitude equal to the input. This square wave can, if required, be fed to an a.c. millivoltmeter, so that very small values of d.c. input voltage can be indirectly measured.

R_1 and $Q1$ are wired as an attenuator network, and $Q1$ is switched on and off by a –ve gate bias applied from the $Q2$-$Q3$ astable multivibrator, which operates at 1 kHz. With an input connected to R_1, and no gate bias applied ($Q2$ on), $Q1$ acts as a very low resistance, so only a negligible voltage appears on $Q1$ drain; with a large –ve gate bias applied ($Q2$ off), $Q1$ acts like a near-infinite resistance, so almost the full input voltage appears on $Q1$ drain. Thus, the output, taken from $Q1$ drain, appears as a square wave with an amplitude proportional to the input. The output should be taken to a fairly high impedance.

When too large a –ve bias is applied to the gate, the gate-to-source junction of $Q1$ starts to avalanche, and a small 'spike' voltage breaks through to the drain, so a small output is obtained even though no d.c. input is connected to the circuit. To prevent this, the circuit must be set up by connecting a d.c. input to the circuit, and then adjusting R_4 until the amplitude of the output just starts to decrease. When set up in this way, avalanching does not occur, and the circuit can reliably be used to chop voltages as low as a fraction of a millivolt.

CHAPTER 3

20 UNIJUNCTION TRANSISTOR PROJECTS

Another important semiconductor device that has been introduced in recent years is the unijunction transistor, or u.j. This is a specialised but very simple device. It uses the symbol shown in Fig. 3.1a, employs the form of construction shown in Fig. 3.1b, and has the equivalent circuit of Fig. 3.1c.

Looking first at Fig. 3.1b, the device is made up of a bar of n-type silicon material with a non-rectifying contact (base 1 and base 2) at either end, and a third, rectifying, contact (emitter) alloyed into the bar part way along its length, to form the only junction within the device (hence the name 'unijunction').

Since base 1 and base 2 are non-rectifying contacts, a resistance appears between these two points, and is that of the silicon bar. This inter-base resistance is given the symbol R_{BB}, normally has a value between 4,000 and 12,000 Ω, and measures the same in either direction.

In use, base 2 is connected to a +ve voltage, and base 1 is taken to ground, so R_{BB} acts as a voltage divider with a gradient varying from maximum at base 2 to zero at base 1. The emitter junction is connected at some point between base 1 and base 2, so some fraction of the base 2 voltage appears between the emitter junction and base 1. This fraction is the most important parameter of the u.j., and is called the 'intrinsic stand-off ratio', η, and usually has a value between 0.45 and 0.8.

The equivalent circuit of Fig. 3.1c illustrates the above points. r_{B1} and r_{B2} represent the resistance of the silicon bar, and diode $D1$ represents the junction formed between the emitter and the bar. When an external voltage, V_{BB}, is applied to base 2, a voltage of $\eta.V_{BB}$ appears across r_{B1} and on the cathode of $D1$. If, under these conditions, a +ve input voltage, V_E, is applied between the emitter and base 1, but is less than $\eta . V_{BB}$, diode $D1$ becomes reverse biased, so no appreciable current flows from emitter to base 1, since the emitter appears as the

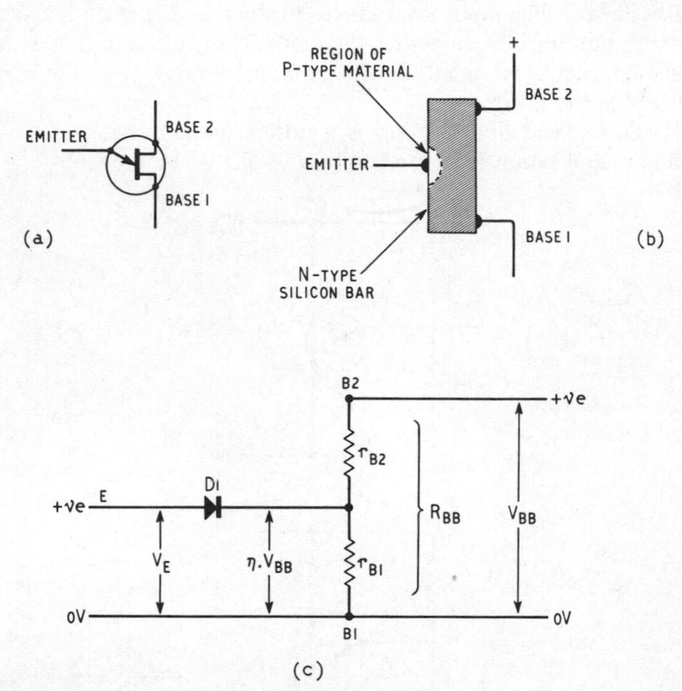

Fig. 3.1

(a) Unijunction symbol. (b) Unijunction construction. (c) Unijunction equivalent circuit

very high impedance of a reverse biased silicon diode, with a typical impedance of several megohms.

If, on the other hand, V_E is steadily increased above $\eta . V_{BB}$, a point is reached where $D1$ starts to become forward biased, so current starts to flow from emitter to base 1. This current consists mainly of minority carriers injected into the silicon bar, and these drift to base 1 and cause a decrease in the effective resistance of r_{B1}; this decrease in r_{B1} causes a decrease in the $D1$ cathode voltage, so $D1$ becomes more heavily forward biased, and the emitter-to-base 1 current increases and in turn causes the r_{B1} value to fall even more. A semi-regenerative action takes place, and the emitter input impedance falls sharply, typically to a value of about 20 Ω.

Thus, the unijunction transistor acts as a voltage-triggered switch and has a very high input impedance (to the emitter) when it is off, and a low input impedance when it is on. The precise point at which triggering occurs is called the 'peak-point' voltage, V_P, and is about 600 mV above $\eta.V_{BB}$.

It can be seen that the u.j. is a rather specialised device. Its most common application is as a relaxation oscillator, as shown in Fig. 3.2a.

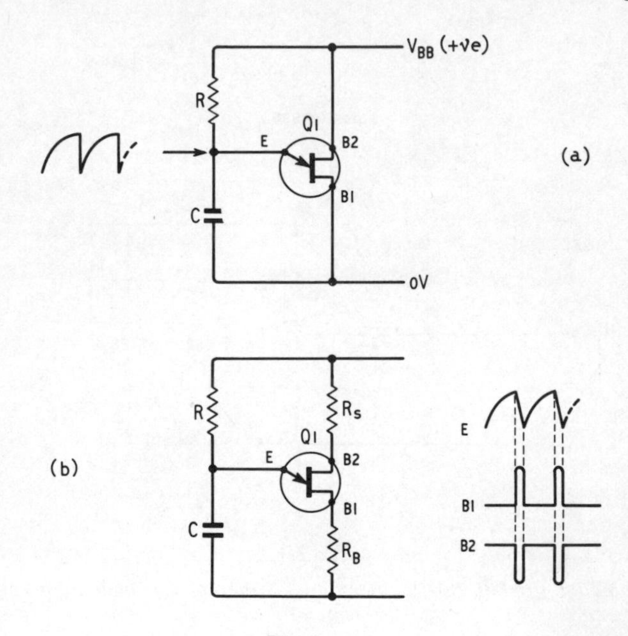

Fig. 3.2

(a) Basic relaxation oscillator. (b) Temperature stabilised relaxation oscillator

Here, when the supply is first connected, C is discharged and the emitter is at ground potential, so the emitter appears as a high impedance; C then charges exponentially towards V_{BB} via R, but as soon as the emitter reaches V_P the u.j. fires and C discharges rapidly into the low impedance of the emitter. Once C is effectively discharged, the u.j. switches off and C starts to charge up again, and the process is repeated. Thus, a rough saw-tooth waveform is continuously generated between $Q1$ emitter and ground.

In this circuit, final switch-off occurs on each cycle when the capacitor discharge current falls to a 'valley-point' value, I_V, typically of several milliamperes. A minimum 'peak-point emitter current', I_P, is needed to switch the u.j. on initially, and typically is a value of several microamperes.

The frequency of operation of the circuit is given approximately by $f = 1/C.R$, and is virtually independent of V_{BB}. Typically, a 10% change in V_{BB} results in a frequency change of less than 1%. The value of R can be varied from about 3 kΩ to 500 kΩ, so an attractive feature of the circuit is that it can be made to cover a frequency range greater than 100:1 via a single variable resistance.

Frequency stability is good with changes in temperature, and is about 0.04%/°C. The main cause of this variation is the change of about −2mV/°C that occurs in the forward volt drop of the $D1$ junction with changes in temperature. Stability can be improved by either wiring a couple of silicon diodes in series with base 2, or by connecting a stabilising resistor, R_S, in the same place. The R_{BB} of the u.j. increases by about 0.8%/°C, so changes in the forward volt drop of $D1$ can be countered by the changes in potential divider action of R_S and R_{BB} with changes in temperature. The correct value of R_S is given by

$$R_S = \frac{0.7\,R_{BB}}{\eta . V_{BB}} + \frac{(1-\eta)R_B}{\eta}$$

where R_B = external load resistor (if any) in series with base 1. An exact R_S value is not important in most applications, however.

In some circuits, R_B is wired between base 1 and ground, as shown in Fig. 3.2b either to control the discharge time of C or to give a +ve output pulse during the discharge period. A −ve pulse is also available across R_S in this period, if needed.

TABLE 3.1

CHARACTERISTICS OF THE 2N2646 UNIJUNCTION TRANSISTOR

Emitter Reverse Volts (max)	= 30 V
V_{BB} (max)	= 35 V
Peak Emitter Current (max)	= 2 A
R.M.S. Emitter Current (max)	= 50 mA
Power Dissipation (max)	= 300 mW
η	= 0.56–0.75
R_{BB}	= 4.7–9.1 kΩ
I_P (max)	= 5 μA
I_V (max)	= 4 mA
case	= T018

Now that the basic principles of the u.j. have been described above, we can select a practical unit and then go on to consider 20 or so

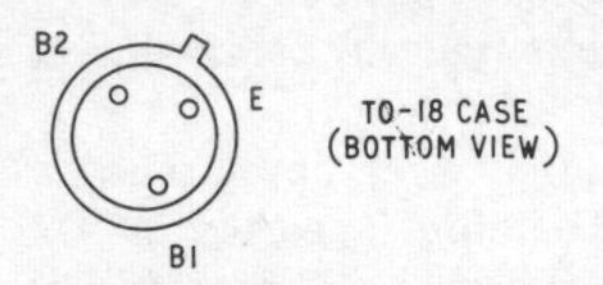

Fig. 3.3
Lead connections of the 2N2646 unijunction transistor

applications in which it can be used. The 2N2646 u.j. has been selected for this purpose, and Table 3.1 and Fig. 3.3 show the general characteristics and lead connections of this particular device.

Wide-range pulse generator

Fig. 3.4 shows the practical circuit of a wide-range pulse generator. A large amplitude +ve pulse is available across R_4, and a –ve pulse across R_3; both pulses have an amplitude of about half supply line volts, are of similar form, and are at a low impedance.

With the component values shown, the pulse width is constant at about 30 μsec over the frequency range 25 Hz–3 kHz (adjustable via R_1). The pulse width and frequency range can be altered by changing

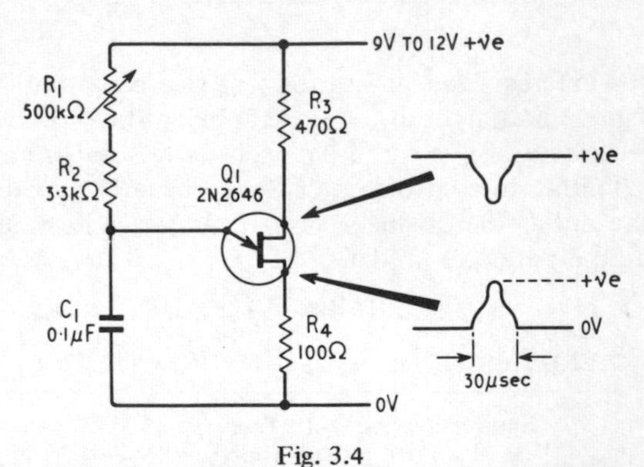

Fig. 3.4
Wide-range pulse generator giving 30 μsec output pulses at repetition frequencies of 25 Hz–3 kHz

the value of C_1. Reducing C_1 by a decade (to 0.01 μF) reduces the pulse width by a factor of 10 (to 3 μsec) and raises the frequency range by a decade (250 Hz–30 kHz). C_1 can have any value in the range 100 pF–1,000 μF.

A saw-tooth waveform is generated at $Q1$ emitter, but is at a high impedance and is thus not readily available externally.

Wide-range saw-tooth generator

In Fig. 3.5 the saw-tooth waveform from $Q1$ emitter is fed to emitter follower $Q2$, making the saw-tooth readily available to external circuits with input impedances greater than about 10 kΩ. If the output is to

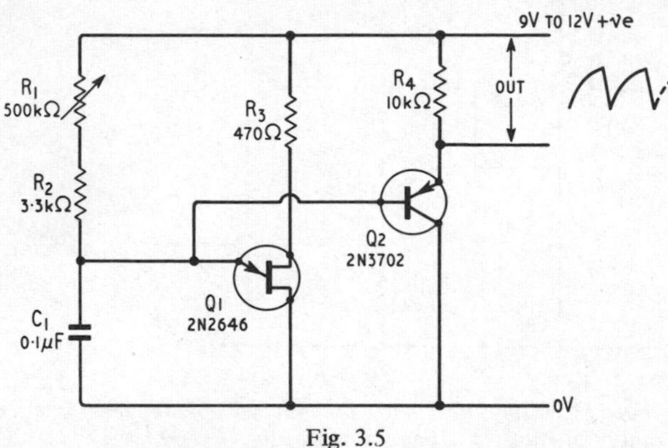

Fig. 3.5

Wide-range saw-tooth generator covering the frequency range 25 Hz–3 kHz

be taken to impedances lower than 10 kΩ, a second emitter follower should be wired between $Q2$ emitter and the output.

With the component values shown, the frequency range of the circuit is variable from about 25 Hz–3 kHz via R_1. The operating frequency can be varied from less than one cycle per minute to over 100 kHz by changing the C_1 value.

Linear saw-tooth generator

The 'saw-tooth' at the emitter of the basic u.j. oscillator is of exponential form. In some applications, however, a perfectly linear saw-tooth is required, and this can be obtained by charging the main timing capacitor from a constant current source, as shown in Fig. 3.6.

In this circuit, $Q1$ is wired as an emitter follower, with emitter load R_4, and feeds its collector current into the main timing capacitor, C_1. The emitter current of $Q1$, and thus the collector current of $Q1$ and the charging current of C_1, is determined solely by the setting of R_2, so the C_1 charging current is constant and this capacitor charges in a linear fashion. Consequently, a linear saw-tooth waveform is generated

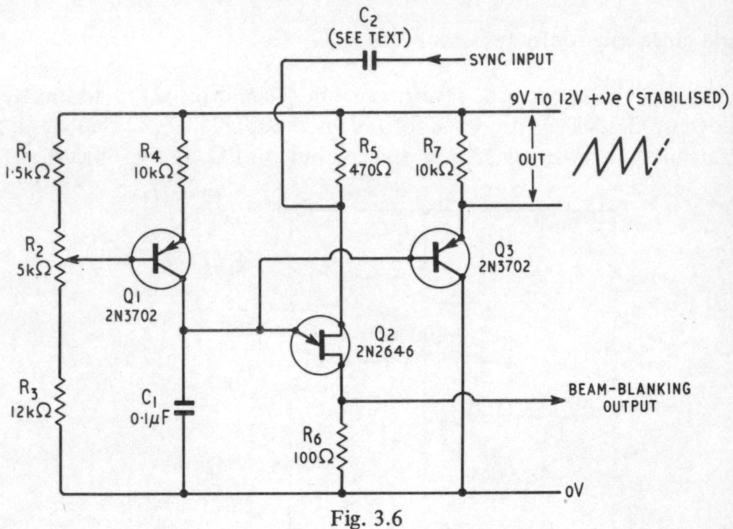

Fig. 3.6

Linear saw-tooth generator, suitable for use as an oscilloscope time-base generator. Frequency range = 50–600 Hz with a 9 V supply, or 70–600 Hz with a 12 V supply

at $Q2$ emitter, and this is made available to external circuits via emitter follower $Q3$. The output should be taken to external circuits with input impedances greater than 10 kΩ.

This particular circuit can be used as a simple time-base generator for an oscilloscope. In this application, the output from $Q3$ emitter should be taken to the external time-base socket of the oscilloscope, and the +ve flyback pulses from R_6 can be taken via a high voltage blocking capacitor and used for beam blanking. The generator can be synchronised to an external signal by feeding the external signal to base 2 of $Q2$, via C_2. This signal, which should have a peak amplitude of between 200 mV and 1 V, effectively modulates the supply voltage, and thus the triggering point, of $Q2$, thus causing $Q2$ to fire in synchrony with the external signal.

C_2 should be chosen to have a lower impedance than R_5 at the sync signal frequency, and should have a working voltage greater than the external voltage from which the signal is applied.

With the components shown, the operating frequency can be varied over the range 50 Hz–600 Hz using a 9 V supply, or 70 Hz–600 Hz using a 12 V supply. Alternative frequencies can be obtained by changing the C_1 value. At very low frequencies, C_1 should be a reversible type of capacitor.

Analogue/digital converter, resistive

The circuit of Fig. 3.7 converts changes in light level, temperature, or any other quantity that can be represented by a resistance, into changes in frequency. The resistive element (l.d.r., thermistor, etc.) is wired in parallel with R_1, and so controls the charging time constant

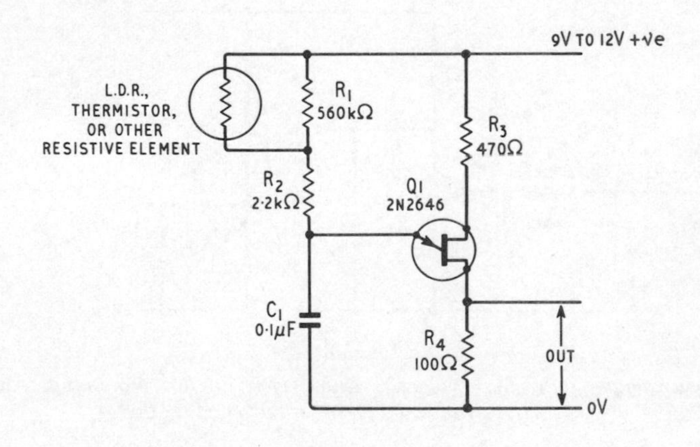

Fig. 3.7
Analogue/digital converter (resistive). With element open circuit, frequency = 30 Hz; with element short circuit, frequency = 3.7 kHz

of C_1, and thus the frequency of operation. A range of 30 Hz–3.7 kHz is available, the lower frequency being obtained with the element open circuit.

The output is taken from across R_4, and consists of a series of pulses. When fed to an earphone, these can be clearly heard, even at the lowest frequency.

The unit is of particular value in remote reading of temperature, etc., the output pulses being used to modulate a radio or similar link. At the receiver end of the link, the digital information can be converted back to analogue via a simple frequency meter type of circuit.

Analogue/digital converters, voltage

The circuits in Figs. 3.8–3.10 have similar applications to the resistance controlled circuit already mentioned, but have their operating frequencies controlled by voltage, or any quantity that can be represented by a voltage, i.e., via photo-voltiac cells, thermocouples, etc.

Fig. 3.8 shows a basic 'shunt controlled' converter. $Q1$ shunts the main timing capacitor, C_1, and so shunts off some of its charging current and effects the operating frequency. If zero voltage is fed to

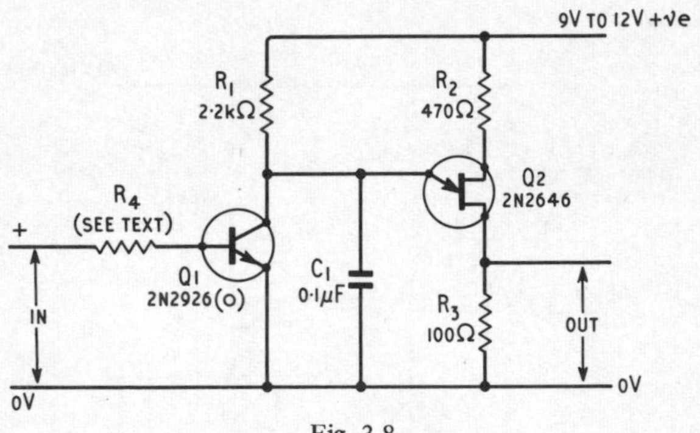

Fig. 3.8

Analogue/digital converter (voltage), shunt type. With zero input voltage, f = 3.7 kHz, with maximum input voltage, f = 800 Hz

$Q1$ base, $Q1$ is cut off, and the circuit operates at maximum frequency (about 3.7 kHz). When a +ve voltage is fed to $Q1$ base, the transistor is driven on, and the operating frequency falls.

A snag with this circuit is that, as $Q1$ is driven on, $Q1$ collector voltage falls, and when it falls to less than V_P, the circuit ceases to operate. The operating range is thus rather restricted, to about 800 Hz minimum in this case.

The value of R_4 is chosen, by trial and error, to suit the control voltage in use, and usually has a value of a few hundred kilohms at potentials up to about 10 V, and a few megohms at 100 V.

Fig. 3.9 shows a basic 'series controlled' converter. Here, the C_1 charging current is controlled almost entirely by $Q1$. When $Q1$ is driven hard on (saturated) by a voltage applied to R_4, the charging current is limited by R_1, and the circuit operates at about 3.7 kHz. When zero voltage is applied to R_4, $Q1$ is cut off, and C_1 charges via R_5, giving an operating frequency of about 30 Hz. Between these two extremes, the

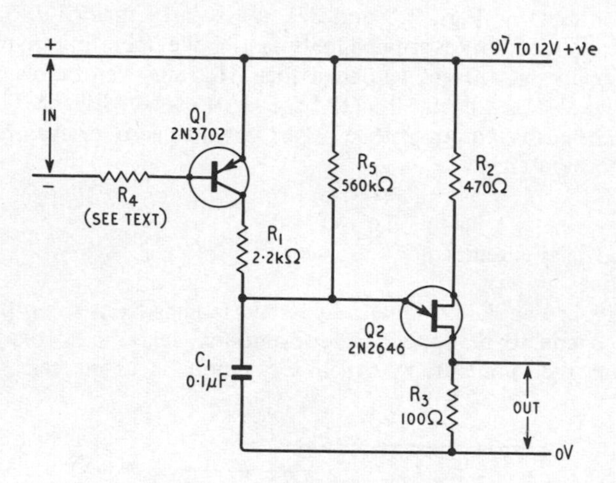

Fig. 3.9

Analogue/digital converter (voltage), series type. With zero input voltage, f = 30 Hz. with maximum input voltage, f = 3.7 kHz

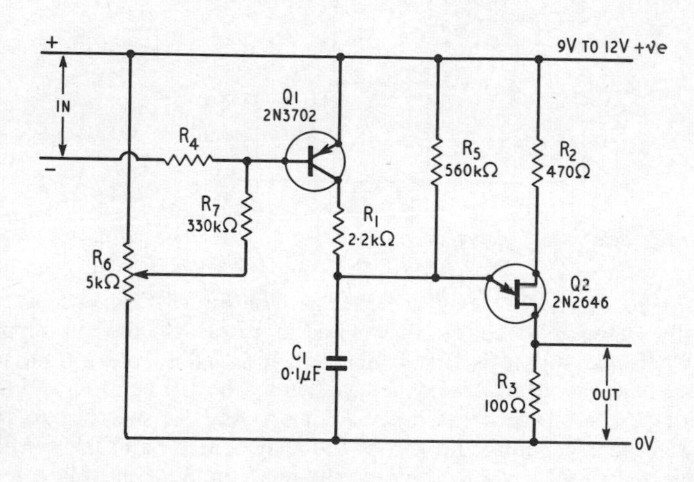

Fig. 3.10

Improved version of Fig. 3.9

frequency can be smoothly controlled by the voltage applied to R_4, and thus be the collector current of *Q*1. The value of R_4 is found by trial and error, to suit individual requirements.

In the circuits of Figs. 3.8 and 3.9, *Q*1 is cut off until a forward voltage of about 650 mV is applied to its base, so the operating frequency is not effected by voltages less than this. This snag can be overcome by applying a standing bias to *Q*1 base, as shown in Fig. 3.10. This modification enables input voltages right down to zero, or even reverse voltages, to be used.

Relay time-delay circuits

The circuits in Fig. 3.11 enable time delays ranging from about 0.5 sec to about 8 min to be applied to conventional relays, i.e., there is a delay from the moment at which the supply is connected to the

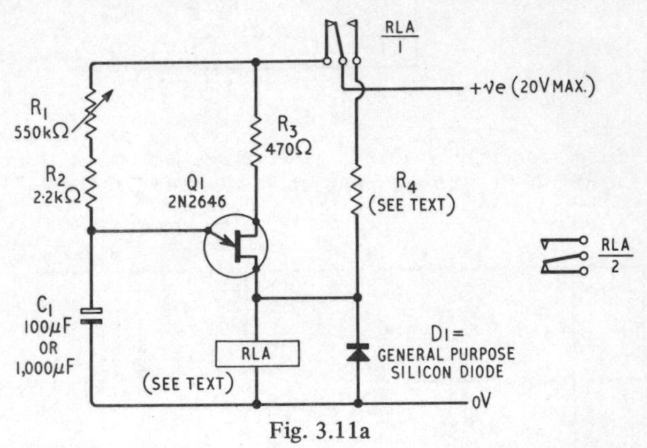

Fig. 3.11a

Basic relay delay unit, giving operating delay of 0.5–50 sec if $C_1 = 100\ \mu F$, *and 3 sec–8 min if* $C_1 = 1{,}000\ \mu F$

moment at which the relay switches on. In Fig. 3.11a, one set of normally closed relay contacts are wired in series with the +ve supply line. When the supply is first connected, it is fed to the u.j. circuit via these contacts. After a delay determined by the setting of R_1 and the value of C_1, the u.j. fires and drives *RLA* on. As *RLA* switches on, the supply to the u.j. circuit is broken by the relay contacts and the +ve line is connected to *RLA* via R_4, holding the relay on. *RLA* must be a fast-acting low-voltage relay with a coil resistance of less than 150 Ω. The supply line potential must be at least 4 times the relay operating voltage, and the value of R_4 must be chosen to keep the 'on' current within limits when the relay is fed from the +ve supply line.

A snag with the circuit of Fig. 3.11a is that the relay type must be carefully selected. This snag is overcome in the circuit of Fig. 3.11b.

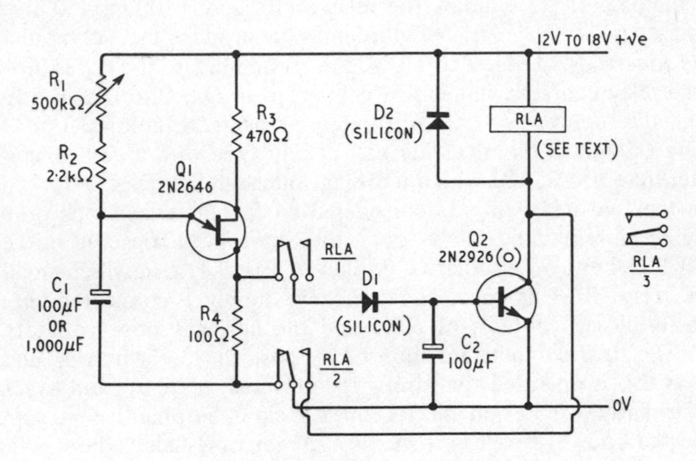

Fig. 3.11b
Alternative relay delay unit, giving same delays as Fig. 3.11a

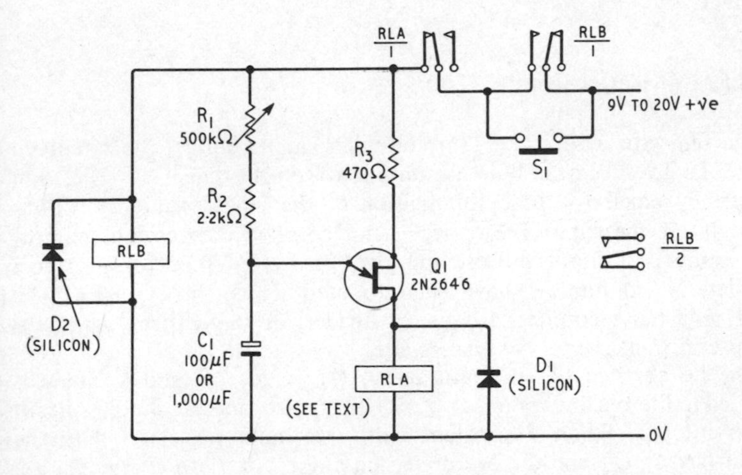

Fig. 3.11c
Current economy version of Fig. 3.11a

Here, the relay is connected to the collector of $Q2$, and is normally off. When the u.j. fires, a +ve pulse is fed from R_4 to the base of $Q2$ via $D1$, driving $Q2$ and *RLA* on, and rapidly charging C_2. At the end of the +ve pulse, the u.j. switches off and $D1$ is reverse biased, so C_2 discharges into the base of $Q2$, holding the relay on for about 100 msec. Thus, C_2 is used as a pulse expander, and eliminates the need for fast-acting relays.

As soon as *RLA* starts to close, the ground line to the u.j. is broken via the relay contacts, but is still connected to $Q2$. Once *RLA* is fully closed, the supply is connected directly across *RLA*, holding it on, and cutting $Q2$ out of circuit. *RLA* can be any type with a coil resistance greater than 100 Ω, and with a working voltage in the range 6–18 V.

In the two relay circuits considered so far, the relays lock on and consume current indefinitely, once they have been triggered initially. Fig. 3.11c shows an alternative version of Fig. 3.11a, in which an additional relay. *RLB*, is used, Here, the +ve supply is connected via the normally closed contacts of *RLA* and the normally open contacts of *RLB*. The *RLB* contacts are shunted by push button switch S_1, and as soon as this is operated the supply is connected to the u.j. and to *RLB*; *RLB* instantly switches on and its contacts close, keeping the +ve supply connected once S_1 is released. After a pre-set time delay, the u.j. fires, driving *RLA* on and thus breaking the +ve supply to the u.j. and to *RLB*, which switches off and thus completely breaks the supply to the circuit. The output of the unit can be taken from the spare *RLB* contacts.

Staircase divider/generator

When fed with a series of constant-width input pulses, the circuit in Fig. 3.12 gives a linear staircase output waveform that has a repetition frequency equal to some sub-division of the input frequency. Alternatively, if the input frequency is not constant, the circuit 'counts' the number of input pulses, and gives an output pulse only after a pre-determined number have been counted. Thus, the circuit can be used as a pulse counter, frequency divider, or step-voltage generator for use in transistor curve tracers, etc.

In the absence of an input pulse, $Q1$ is cut off and $Q2$ base is shorted to the +ve line via R_3, so $Q2$ is cut off also, and no charge current flows into C_2. When a constant-width +ve input pulse is fed to the circuit via C_1, $Q1$ and $Q2$ are driven on and C_2 starts to charge via the collector current of $Q2$, which is wired in the emitter follower mode and acts as a constant current generator, with its collector current controlled via R_6. C_2 charges linearly as long as $Q2$ is on, and since $Q2$ is on only for the fixed duration of the input pulse, the C_2 voltage increases by only a fixed amount each time a pulse is applied. In the absence of the pulse, there is no discharge path for C_2, so the charge voltage stays on this capacitor. The following input pulse again increases the C_2 charge voltage by a fixed amount, until, after a pre-determined number of pulses, the C_2 voltage reaches the trigger potential of $Q3$, and the u.j. fires, discharging C_2 and re-starting the counting cycle.

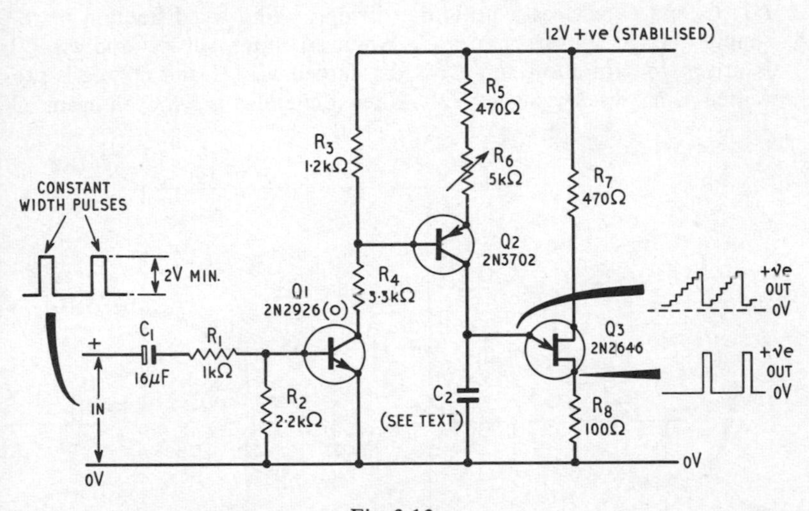

Fig. 3.12

Staircase divider/generator

If the input pulses are applied at a constant repetition frequency, the signal across C_2 is a linear staircase waveform, and an output pulse is available across R_8 every time the u.j. fires. If the input frequency is not constant, the staircase is non-linear, but the R_8 pulse again appears after a pre-determined number of input pulses have been applied. Stable count or division ratios from 1 up to about 20 can be obtained.

It is important to note that this circuit must be fed with constant-width input pulses if stable operation is to be obtained, and that the width of the pulses must be small relative to the pulse repetition period. The value of C_2 is determined by these considerations, and is best found by trial and error. Once a C_2 value has been selected, the division ratio can be varied over a range of about 10:1 via R_6.

Diode-pump counter

The circuit in Fig. 3.13 also acts as a frequency divider or counter, but gives a non-linear staircase output. It has the advantage, however, that counting is almost independent of the shape of the input waveform.

With no input applied, $Q1$ is cut off and C_3 charges via R_3, C_2, and $D1$; C_2 and C_3 acts as a potential divider, and a fixed fraction of the supply voltage appears across C_3. When an input pulse is applied, $Q1$ is driven to saturation and C_2 is discharged via $Q1$ and $D2$; C_3 is prevented from discharging by $D1$. When the pulse is removed again, C_2

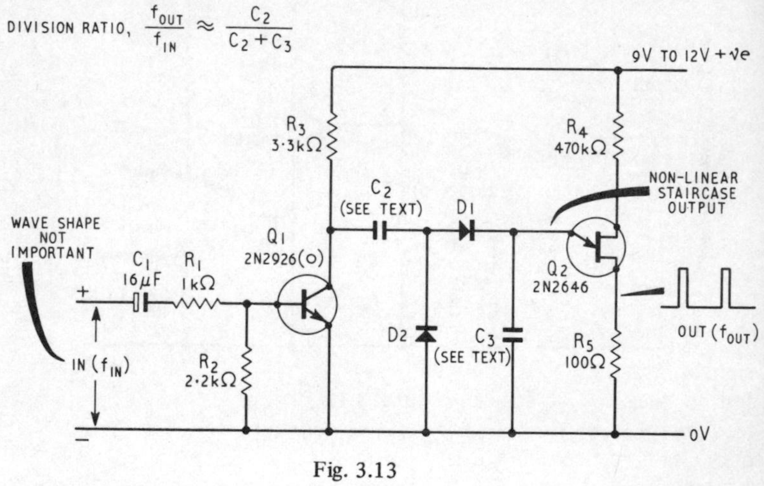

Fig. 3.13

Diode pump counter. N.B. D1 *and* D2 *are general purpose germanium diodes*

again charges via $D1$ and C_3, and places another fraction of the supply voltage on C_3. Thus, at the end of each pulse, the C_3 voltage increases by a fixed step, until eventually the u.j. fires, discharges C_3, and the count cycle starts over again. Pulse shape has virtually no effect on circuit operation.

The division ratio, f_{out}/f_{in}, is roughly equal to $C_2/(C_2+C_3)$. The ratio is, however, effected by a number of variable factors, including operating frequency, so the values of these two components are best found by trial and error. Once component values have been selected, the circuit will give stable division over quite a wide range of input frequency variation. Stable division ratios up to 10:1 can be easily obtained.

Synchronised frequency divider

The circuit in Fig. 3.14 is useful in generating standard timing waveforms or frequency standards. Positive pulses from a 100 kHz crystal oscillator are fed, via C_1, to base 2 of $Q1$, and R_1 is adjusted so that the u.j. locks firmly to an operating frequency of 10 kHz, the 100 kHz

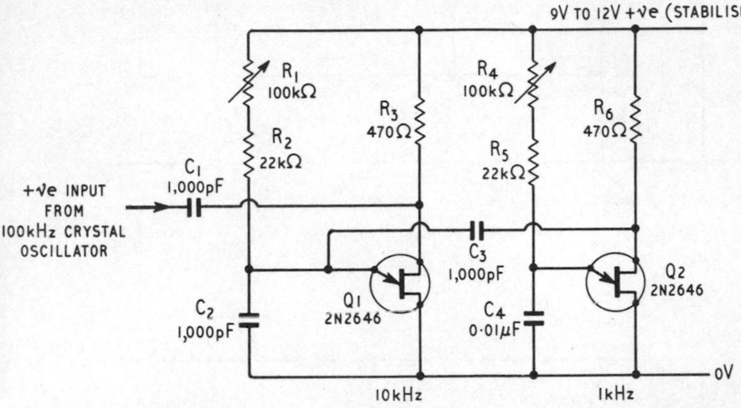

Fig. 3.14
Synchronised frequency divider, giving standard frequencies (and times) of 100 kHz (10 μsec), 10 kHz (100 μsec), and 1 kHz (1 msec)

signals acting as sync pulses. The 10 kHz signal from $Q1$ emitter is fed to $Q2$ via C_3, and R_4 is adjusted so that $Q2$ locks to an operating frequency of 1 kHz. Thus, the circuit makes available standard frequencies (and times) of 100 kHz (10 μsec), 10 kHz (100 μsec), and 1 kHz (1 msec). Stability is excellent if a zener stabilised supply line is used.

Division ratios other than 10 can be obtained by adjusting R_1 and R_4. Outputs can be taken, via a high impedance emitter follower buffer stage, from the emitter of each u.j. and from the crystal oscillator.

Wide-range square wave generators

The u.j. can be used as the basis of a whole range of different waveform generators. Figs. 3.15a and b show how it can be used to generate square waves.

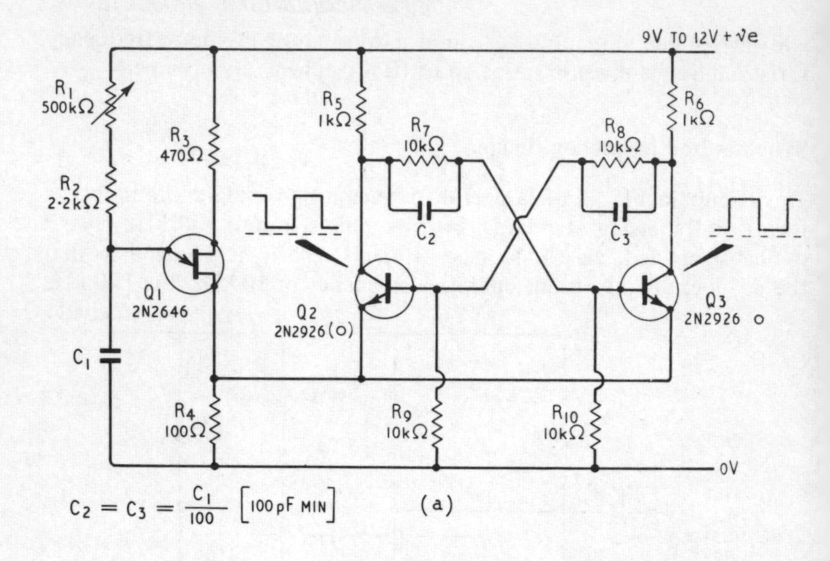

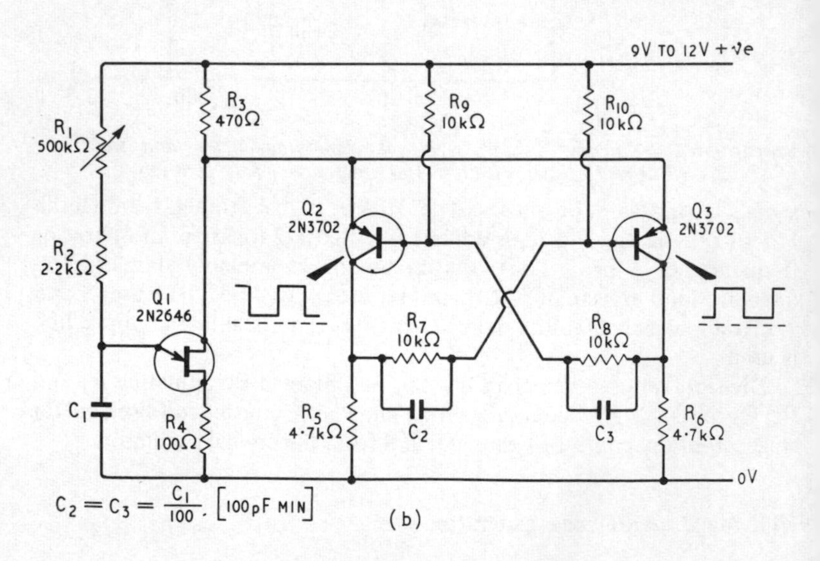

Fig. 3.15

(a) Wide-range square wave generator (npn) C_1, C_2, *and* C_3 *are selected to suit frequency range required. (b) pnp version of the wide-range square wave generator*

In Fig. 3.15a, *Q*2 and *Q*3 form an npn bistable multivibrator or divide-by-two circuit. At the end of each u.j. cycle, the +ve pulse from R_4 is fed to the emitters of *Q*2 and *Q*3 and cause the multivibrator to change state. Two cycles of the u.j. result in a single complete cycle of the multivibrator, so the multivibrator output, taken from either collector, is a perfect square wave at half of the u.j. frequency. The two collector signals are in anti-phase.

Fig. 3.15b shows the pnp version of the same circuit. In this case, the circuit uses the –ve pulses from R_3 to trigger the bistable multivibrator, but the two circuits are otherwise similar.

It's important to note that in both of these circuits C_2 and C_3 are of equal value, and have a value of approximately $C_1/100$, i.e., if C_1 = 0.1 μF, C_2 and C_3 = 0.001 μF (= 1,000 pF). C_2 and C_3 should, however, have a maximum value of about 100 pF.

Both Fig. 3.15a and b will generate square waves over a 100:1 frequency range, using a single set of component values.

Variable frequency pulse generator

The circuit in Fig. 3.16 generates a constant-width pulse that can be varied in repetition frequency over a 100:1 range. It may, for

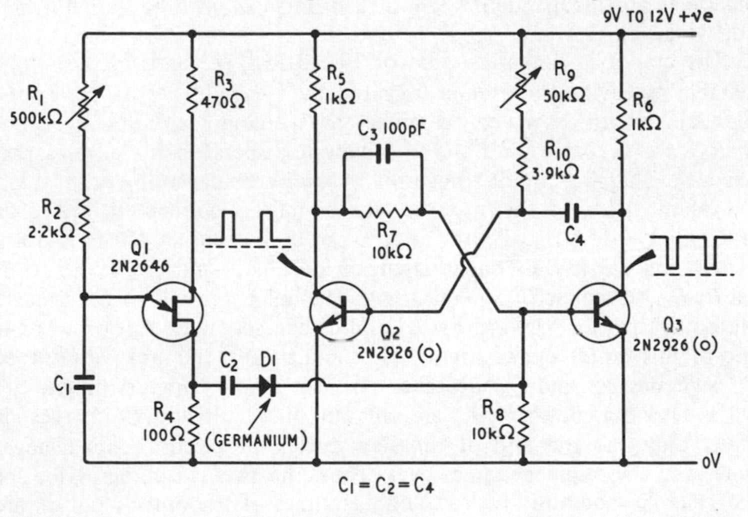

Fig. 3.16

Variable frequency pulse generator

example, generate a pulse with a constant width of 500 μsec, at repetition frequencies ranging from 10 Hz to 1 kHz. The actual pulse width can be adjusted, on any particular range, over a 10:1 range, i.e., from 50–500 μsec.

The circuit is quite simple. $Q2$ and $Q3$ are wired as a monostable or one-shot multivibrator, with pulse width controlled by R_9-R_{10} and C_4, and the multivibrator is triggered by the +ve pulses fed from R_4 to $Q3$ base via C_2 and $D1$. Thus, repetition frequency is controlled by the u.j. and pulse width by the multivibrator.

Different sets of C_1-C_2-C_4 values are needed for each range of operation, but all three capacitors are usually of equal value. The main point here is that the maximum period of the pulse must be less than the minimum period of the u.j. cycle, otherwise the pulse will not be ended by the time a new trigger pulse arrives, and stable operation will not be obtained.

Pulse outputs can be taken from either collector, the two outputs being in anti-phase.

Variable on/off-time pulse generator

The circuit in Fig. 3.17 generates a series of pulses in which the on and off times are independently controlled and can each be varied over a 100:1 range.

The circuit is similar to that of Fig. 3.15a, $Q2$ and $Q3$ forming a bistable multivibrator that is triggered by +ve pulses from R_6. In the Fig. 3.17 circuit, however, two different C_1 charging circuits (R_1-R_2 and R_3-R_4) are available, and the multivibrator operates diode gates that select the charging circuit to be used at any particular moment.

Assume that, at the moment the supply is connected, $Q2$ is on and $Q3$ is off. $Q2$ collector is near ground volts, so $D4$ is forward biased and $D3$ is thus back-biased, so no charge current flows to C_1 via R_3-R_4. $Q3$ collector is at near full +ve rail potential, so $D2$ is back-biased; $D1$ is thus forward biased and C_1 charges via R_1-R_2 only. At the end of this timing cycle, the u.j. fires and triggers the multivibrator, so $Q2$ switches off and $Q3$ switches on. $D2$ is now forward biased and $D4$ is back-biased, so R_1-R_2 are cut out of circuit and C_1 charges via R_3-R_4 only. At the end of this new cycle, the circuit again changes state, and the sequence starts over. Thus, the two switching periods of the bistable, and thus the on and off times of the output pulses, are individually controlled.

C_2 and C_3 are of equal value and = $C_1/100$, down to a minimum of

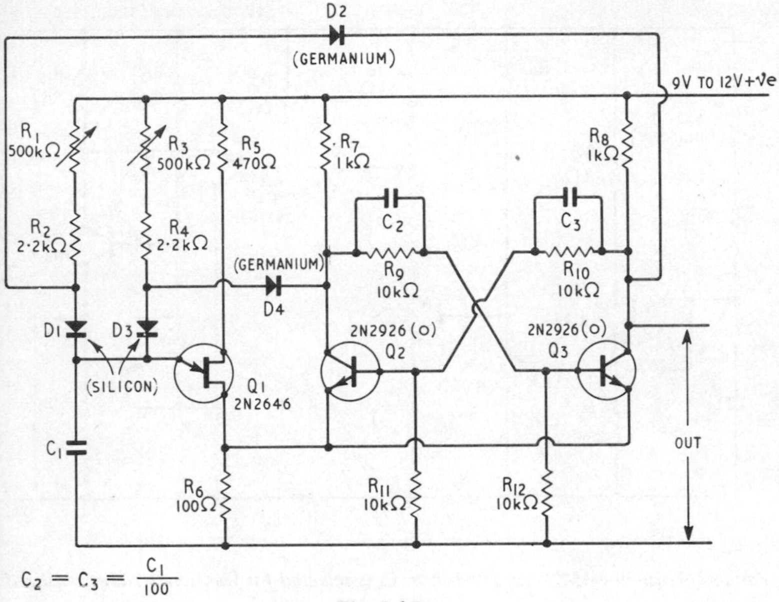

Fig. 3.17

Variable on/off-time pulse generator. With C_1 = 0.1 μF, individual on and off times are variable from 500 μsec–50 msec

100 pF. With C_1 = 0.1 μF, the on and off times can be individually controlled over the range 500 μsec to 50 msec.

Variable frequency/M-S ratio generator

The circuit in Fig. 3.18 generates a series of pulses in which both the mark-space ratio and the frequency can be independently varied over a wide range. If, for example, the M-S ratio is set at 9:1, the operating frequency can be varied from (say) 10 Hz to 1 kHz without any resulting change in M-S ratio. Similarly, if the frequency is set at (say) 100 Hz, the M-S ratio can be varied over the range 1:100 to 100:1 without any resulting change in operating frequency. Both frequency and M-S ratio can be simultaneously varied, without interaction. This type of generator is often used at the transmitter end of analogue dual-proportional radio control systems, such as 'Galloping Ghost'.

In Fig. 3.18, *Q*2 and *Q*3 form a Super-Alpha pair emitter follower, and enable a saw-tooth output to be taken at low impedance from the

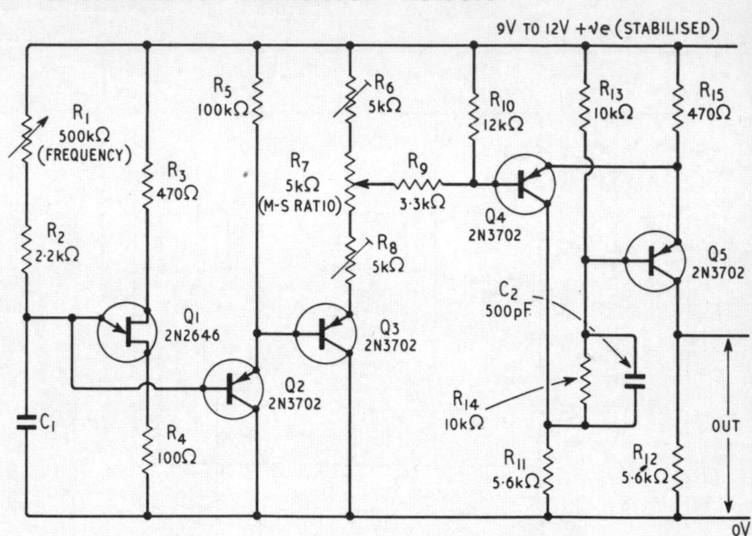

Fig. 3.18

Variable frequency/M-S ratio generator. C_1 *is selected for frequency range required*

R_6-R_7-R_8 chain without effecting the operating frequency of *Q*1. This saw-tooth is then fed, via R_9, to the Schmitt trigger formed by *Q*4 and *Q*5, and by adjusting R_7 the Schmitt can be made to fire at different points on the saw-tooth, and so generate different M-S ratio pulse signals at *Q*5 collector. R_6 and R_8 enable the maximum and minimum M-S ratios to be pre-set. Frequency ranges can be selected via C_1, as in all u.j. circuits, and, in any given range, the frequency can be varied via R_1. Thus, R_7 acts as the M-S ratio control, and R_1 as the frequency control.

One-shot lamp/relay driver

Fig. 3.19 shows the circuit of a one-shot lamp or relay driver. Here, the lamp or relay is normally off, but comes on as soon as push button S_1 is operated, and then stays on for a pre-set period that can be adjusted from about 4 sec to 8 min. At the end of this period, the lamp or relay switches off and the circuit re-sets, ready for the next operation of S_1.

*Q*2 and *Q*3 form a bistable multivibrator, in which *Q*2 is normally

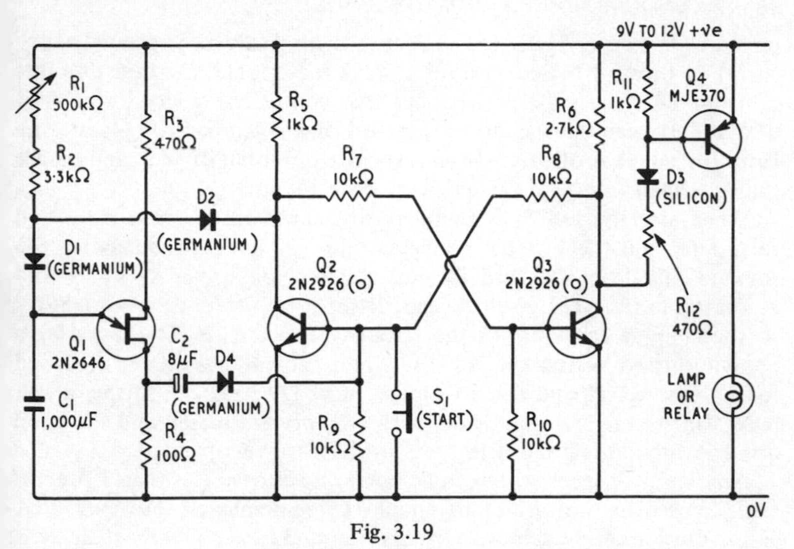

Fig. 3.19

One-shot lamp/relay driver, giving 'on' times variable from 4 sec–8 min

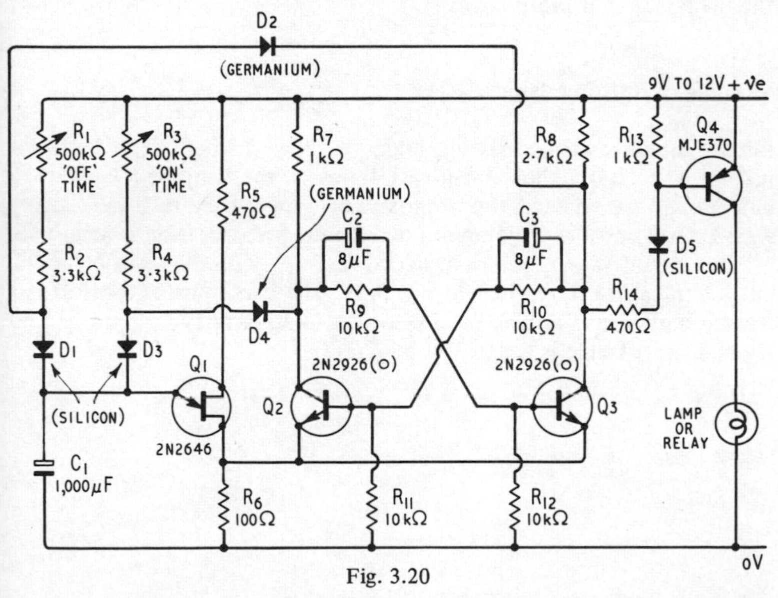

Fig. 3.20

Variable on/off-time lamp flasher, giving 'on' and 'off' times individually variable from 4 sec–8 min (= 16 min maximum total)

on and $Q3$ is off. Thus, $Q2$ collector is normally near ground volts, so $D2$ is forward biased and $D1$ is back biased, and $D1$ thus prevents C_1 from charging via R_1-R_2. Under this condition, $Q3$ collector is at near full +ve rail voltage, so no forward bias is applied to $Q4$, and the lamp (or relay) is off; (R_{11}-$D3$-R_{12} form a potential divider, and ensure that the small voltage at $Q3$ collector does not turn $Q4$ on).

When start button S_1 is momentarily operated, $Q2$ base is shorted to ground and the bistable changes state. $Q2$ goes off, removing the forward bias from $D2$, and C_1 thus starts charging via R_1-R_2-$D1$, and at the same time $Q3$ goes on and drives $Q4$ to saturation via $D3$-R_{12}, so the lamp switches on. C_1 then charges up via R_1-R_2-$D1$, and after a pre-determined period the u.j. fires, and the +ve pulse from R_4 is fed to $Q2$ base via C_2 and $D4$; this pulse turns $Q2$ back on, so the circuit re-sets in its original condition, with $D2$ forward biased and the lamp off. The circuit maintains this state until S_1 is again operated.

Any lamp or relay with a peak operating current less than 1 A or so can be used in this circuit. It should be remembered, however, that lamps draw peak switch-on currents about 3 times greater than their normal running currents. Alternative silicon transistors can be used in the $Q4$ position, if preferred.

Variable on/off-time lamp flasher

Finally, another sequential u.j. lamp or relay driving circuit is shown in Fig. 3.20. Here, the on and off times of the lamp or relay can be individually varied over the approximate range 4 sec to 8 min, giving a maximum possible cycle period of 16 min, and operation is repetitive.

This circuit is simply a re-hash of Fig. 3.17, with the addition of the output transistor stage given in Fig. 3.19. The maximum output current is again limited to 1 A. The on time of the lamp or relay is controlled by R_3, and the off time is controlled by R_1.

CHAPTER 4

15 SILICON CONTROLLED-RECTIFIER PROJECTS

The three types of semiconductor device that we have looked at so far have been developed primarily for low power applications. New types of device have also been developed for use in high power switching circuits, however, and one of the most important of these is the Silicon Controlled-Rectifier, or *SCR*, which is also known as a Thyristor.

The *SCR* uses the symbol shown in Fig. 4.1a. Note that this symbol resembles that of a normal rectifier, but has an extra terminal, known as the 'gate'. The *SCR* should, in fact, be regarded as a modified silicon rectifier, giving the following basic characteristics:

(1) Normally, with no bias applied to the gate, the *SCR* is 'blocked', and acts, between the anode and the cathode, like an open circuit switch; it passes negligible current in either direction.

(2) When a suitable +ve bias is fed to the gate, the *SCR* acts like a normal silicon rectifier, and conducts (between anode and cathode) in the forward direction, but blocks in the reverse direction.

(3) Once the *SCR* has turned on and is conducting in the forward direction, the gate loses control, and the *SCR* stays on even though the gate bias may be removed. Thus, only a brief +ve gate pulse is needed to turn on the *SCR*.

(4) Once the *SCR* has turned on, it can only be turned off again by reducing its internal currents to zero. In a.c. circuits, turn-off thus occurs automatically on the –ve half of each cycle. The *SCR* can *not* be turned off via the gate.

In use, an external load is wired in series with the *SCR*, which is then operated as a switch. This mode of operation enables the device to switch high power loads with high efficiency. Suppose, for example, that the *SCR* is wired to a load into which it is required to switch 3 A from a 400 V supply. With the *SCR* blocked-off only small leakage currents flow, so negligible power is dissipated in the circuit, but when the *SCR* is switched on it passes the full 3 A through itself and the load; only about 2 V are developed across the *SCR* when it is on, however, so only 6 W are developed in the *SCR*, while nearly 1,200 W are developed in the external load.

A major advantage of the *SCR* is that it offers a high power gain between the gate and external load. Typically, a maximum gate current of 20 mA at 2 V is needed to trigger a 3 A *SCR*, so, in the above example, the overall power gain is 30,000.

*SCR*s can be used to replace conventional relays. They have no moving parts to wear out or arc, are silent in operation, can operate at high speeds, and are not adversely affected by severe mechanical vibration or by high 'G' forces.

SCR theory

The *SCR* is a four-layer npnp device. Fig. 4.1b shows a simplified diagram of its structure, while Fig. 4.1c shows an alternative representation. From this second diagram it can be seen that the *SCR* can be roughly simulated by an npn and pnp transistor connected as shown in Fig. 4.1d, and that *SCR* operation can thus be explained in transistor terms. R_1 and R_2 represent the semiconductor resistance between gate and cathode; circuit action is explained as follows:

When the supply is first connected, and with zero bias on the gate, $Q1$ base is shorted to the cathode via R_1 and R_2, so $Q1$ is cut off and passes no collector current. $Q2$ base current is derived from $Q1$ collector, so $Q2$ is also cut off under this condition, and zero current flows between anode and cathode.

When, on the other hand, a +ve bias is applied to the gate, $Q1$ is driven on. The resulting collector current of $Q1$ feeds directly into the base of $Q2$, and drives that transistor on also. The resulting collector current of $Q2$ feeds back into the base of $Q1$, thus completing a positive feedback loop. Regenerative action takes place, and both transistors are driven to saturation, and a heavy current flows between anode and cathode. Once regeneration starts, it continues independently of the applied gate voltage, and once both transistors are saturated they can only be turned off again by momentarily reducing

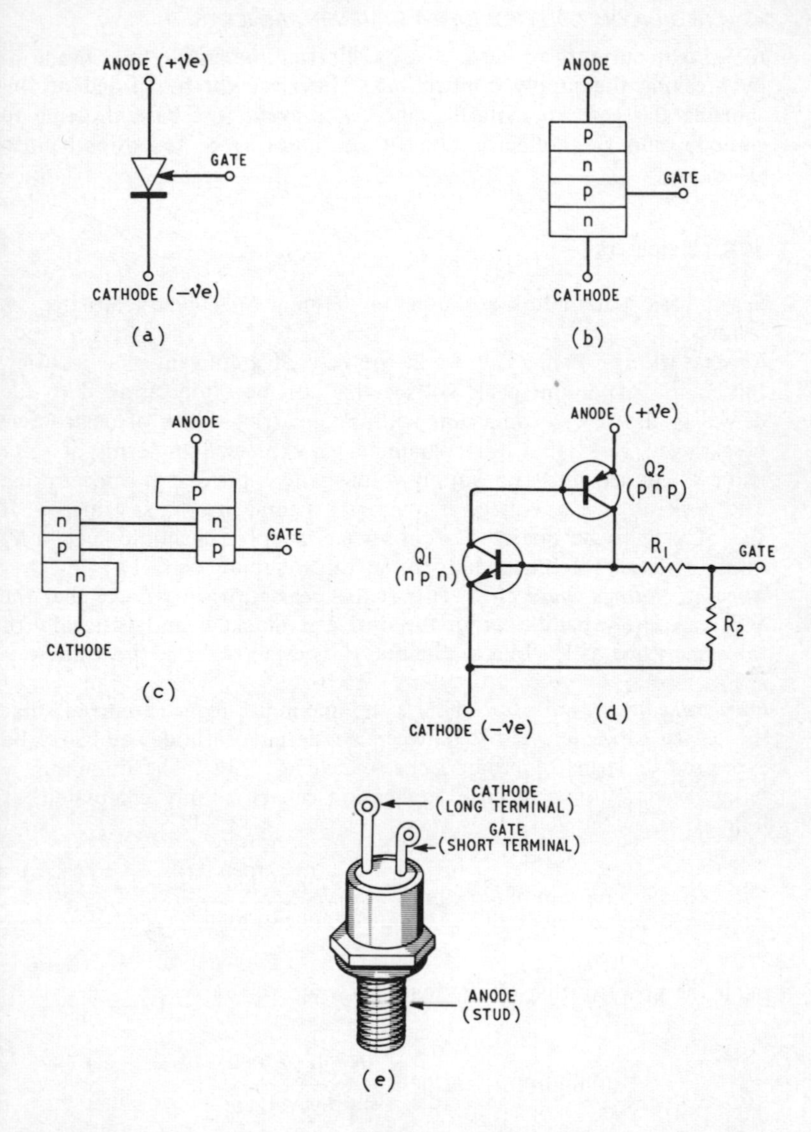

Fig. 4.1

(a) Symbol for silicon controlled-rectifier, or thyristor. (b) Simplified diagram of SCR *semiconductor structure. (c) Alternative representation of* SCR *semiconductor structure. (d) Simple transistor equivalent of* SCR, *derived from Fig. 4.1c. (e) Connections of* SCR *using stud type of construction*

the circuit currents to zero, i.e., by shorting the anode to cathode or by breaking the supply connections. They can *not* be turned off by shorting the gate to cathode, since R_1 prevents $Q1$ base shorting to cathode, and $Q2$ collector current continues to be fed directly into $Q1$ base.

SCR parameters

Seven basic parameters are used in defining *SCR* characteristics, as follows:

Reverse voltage, max (V_r). As in the case of a conventional rectifier, this is the maximum peak voltage that can be safely applied to the device in the reverse direction without incurring a risk of destructive breakdown. Note that this parameter is expressed in terms of *peak* voltage, whereas most a.c. supply voltages are expressed in r.m.s. rating. The peak of an a.c. voltage is roughly 1.4 times its r.m.s. value, so, if the *SCR* is to be operated from an a.c. supply, it should have a V_r rating at least 1.4 times that of the r.m.s. supply voltage.

Forward voltage, max (V_f). This is the peak forward voltage that the *SCR* can safely handle when the device is blocked, and is usually of the same value as V_r. In a.c. circuits, V_f is established in the same way as V_r.

Forward current, max (I_f). This is the maximum forward current that the device can safely carry between anode and cathode, and may be expressed in terms of either r.m.s. or average value. The minimum I_f rating needed for a specific application can be simply calculated, as follows:

$$\text{minimum } I_f \text{ rating} = \frac{\text{supply voltage}}{\text{resistance of load}}.$$

or, if the power rating of the load is known:

$$\text{minimum } I_f \text{ rating} = \frac{\text{power of load}}{\text{supply voltage}}.$$

Thus, if an *SCR* is required to control a maximum load of 1 kW from a 230 V a.c. supply, the *minimum* I_f rating = 1,000/230 = 4.35 $A_{r.m.s.}$ When making these calculations it should be borne in mind that electric fires and lamps may, at the moment of switch-on, dissipate three times their normal running power, while electric drills may dissipate several times their normal running power when stalled or heavily loaded.

Gate voltage, max to trigger (V_g). This is the maximum forward gate bias voltage needed to trigger the *SCR,* and typically has a value of 1-2 V.

Gate current, max to trigger (I_g). The maximum gate current, I_g, required to trigger the *SCR* typically has a value between 1 and 30 mA. The gate-to-cathode junction acts like a normal silicon diode, and presents a very low impedance when forward biased, so in practical circuits the gate current should be limited to some safe value above I_g via a series resistor, which should have a maximum value selected on the basis of:

$$\text{maximum resistance} = \frac{\text{gate voltage}}{Ig}.$$

The maximum permissible gate current of the *SCR* is usually limited to about a tenth of I_f, so the minimum value of the gate resistor can be calculated from:

$$\text{minimum resistance} = \frac{10 \times \text{gate voltage}}{I_f}.$$

The final value of gate resistor should rest between these two extremes.

Holding current, max (I_{hm}). It was mentioned earlier that, to turn off an *SCR,* its currents must be reduced to zero. In practice, however, it is usually possible to turn off the device by simply reducing the currents to a fairly low value, typically between 1 and 50 mA. Consequently, the *SCR* may not hold on correctly if operated with too low an anode current, and a minimum holding current, I_h, is therefore specified in manufacturers data sheets. I_{hm} is the maximum value of I_h occuring in a production spread of *SCR*s, and its practical effects are to limit the maximum resistive values of anode load that can be reliably used.

Peak on-voltage drop at I_f (V_{fm}). This is the maximum forward voltage drop of the *SCR* when operating at maximum current rating, and typically has a value of about 2 V.

This completes the description of the general characteristics of the *SCR*, and we can now go on to look at 15 practical circuits of interest to the experimenter. All of these circuits are intended for low-voltage work, and have been designed to work with *any SCR* with an I_f of 3 $A_{r.m.s.}$ and a V_f of 50 V, so any *SCR* meeting or exceeding these requirements can be used. Most *SCR*s use a stud type of construction, and Fig. 4.1e shows the usual connections.

Basic d.c. on/off circuits

Fig. 4.2 shows a basic *SCR* d.c. on/off circuit, driving a 12 V, 500 mA lamp load. Any type of load drawing a maximum current less than 3 A can, in fact, be used here, but the *SCR* may need to be mounted on a heat sink at currents above 1 A or so. If an inductive load is used, it

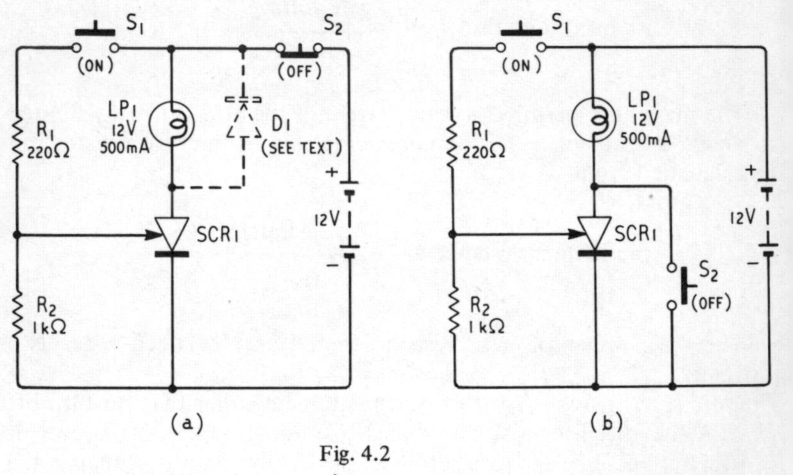

Fig. 4.2

(a) Basic SCR *d.c. on/off circuit*

(b) Alternative on/off circuit

In both circuits, SCR*1 is any* SCR *with a* V_r *of 50 p.i.v. and an* I_f *of 3 A, or greater*

must be shunted by a reverse connected diode, with a current rating equal to that of the load, as shown dotted in the diagram, to prevent high back e.m.f.s damaging the circuit.

The *SCR* and lamp are turned on by briefly connecting a +ve gate voltage via push-button S_1. The circuit is self-latching, and the gate bias only has to be applied for a couple of microseconds to ensure full turn-on. S_1 can be omitted if preferred, and the turn-on gate pulse can be applied via a transistor pulse generator. The *SCR* is turned off by momentarily breaking the supply connections via S_2; the *SCR* takes a few tens of microseconds to turn off fully.

An alternative method of turning off the *SCR* is shown in Fig. 4.2b. Here, the *SCR* anode is shorted to the cathode when S_2 is momentarily operated, so the *SCR* currents are briefly reduced to zero and switch-off again occurs.

A variation of this switch-off theme is shown in Fig. 4.3. Here, with the *SCR* on, C_1 charges via R_3. When fully charged, the *SCR*-anode end of C_1 is 2 V above ground potential, and the R_3 end is at full +ve rail

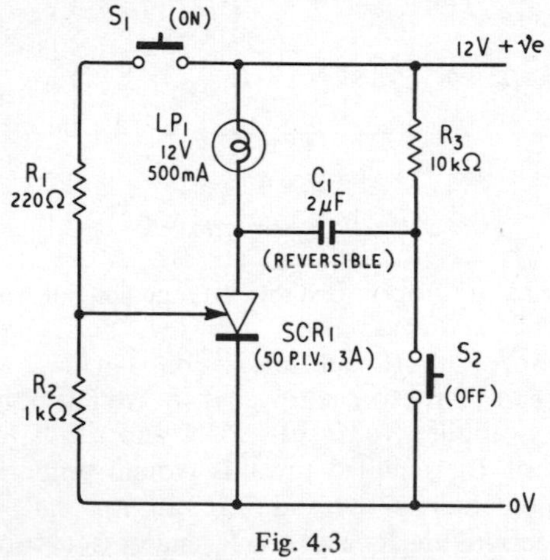

Fig. 4.3
Capacitor–discharge turn-off circuit

voltage, giving a capacitor charge of 10 V in this particular case. Now, when S_2 is operated, the +ve end of C_1 is clamped to ground, and the capacitor charge therefore forces the anode of the *SCR* to momentarily swing to about 10 V –ve, thereby reverse biasing the *SCR* and thus causing it to cut off. The capacitor charge leaks away rapidly under this condition, but only has to hold the *SCR* anode negative for a few hundredths of a millisecond to ensure complete switch off. Note that, if S_2 is held down after the charge has leaked away, the capacitor then starts to charge in the reverse direction via LP_1, so C_1 must be a reversible type. The value of C_1 is not critical.

Fig. 4.4 shows a modification of Fig. 4.3, using an additional *SCR* to enable switch-off to be obtained via a low-current gate pulse. *SCR*1

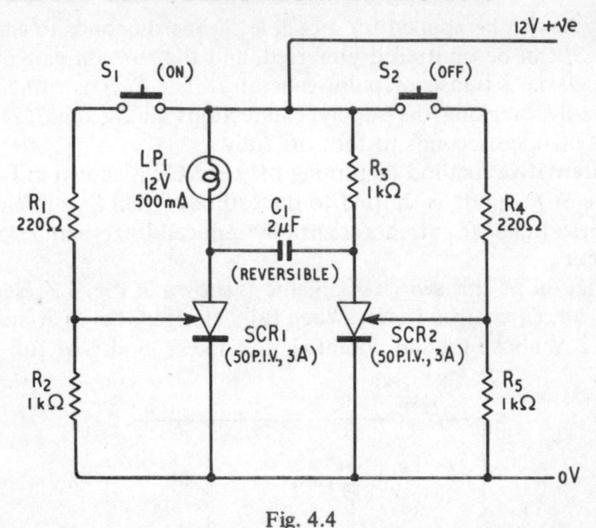

Fig. 4.4

*Dual-*SCR *on/off circuit (bistable)*

and *SCR*2 work as a flip-flop or bistable arrangement, in which *SCR*1 is on when *SCR*2 is off, and vice versa.

Suppose that *SCR*1 is off and *SCR*2 is on. C_1 charges via LP_1, and the *SCR*1-anode end of the capacitor goes to +ve rail potential. When a +ve gate pulse is applied to *SCR*1, *SCR*1 and the lamp go on; the *SCR*1-anode end of C_1 is pulled towards ground potential, so *SCR*2 anode is driven momentarily –ve and *SCR*2 turns off. C_1 then charges in the reverse direction, via R_3, and the R_3-end of C_1 eventually reaches the +ve supply rail potential. Thus, when a +ve gate pulse is applied to *SCR*2, *SCR*2 switches on and pulls the R_3 end of C_1 to near ground potential and so drives *SCR*1 anode –ve, and thereby causes *SCR*1 to switch off. The cycle then repeats *ad infinitum.* In this circuit, *SCR*2 only has to carry a current of V_{supply}/R_3.

Automatic turn-off circuit

Fig. 4.5 shows a development of Fig. 4.4, in which, once the lamp has been turned on via S_1, turn-off occurs automatically after a pre-set period. The turn-off delay is determined by a u.j. timer circuit, and can be varied from about 8–80 sec via R_7.

Normally, *SCR*1 and the lamp are off, and *SCR*2 is on and its

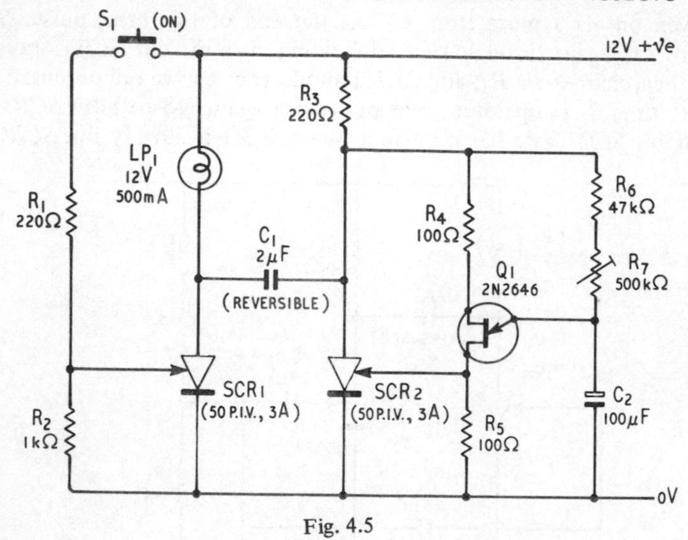

Fig. 4.5

Automatic turn-off circuit, giving switch-off delay of 8–80 sec

anode is at near ground potential. The u.j. circuit is therefore inoperative. When a +ve trigger pulse is applied to *SCR*1, *SCR*1 and the lamp go on and *SCR*2 goes off. As *SCR*2 switches off, its anode rises towards the +ve rail voltage, and the u.j. circuit then starts a timing cycle. At the end of a period determined by the setting of R_7, the u.j. fires and triggers *SCR*2 on via a +ve pulse from R_5, and *SCR*2 triggers *SCR*1 off via C_1. The circuit is thus re-set and ready for the next operation of S_1.

Note that, when the supply is first connected, both *SCR*s are off, so there is a delay in which the u.j. goes through one complete cycle before the circuit takes up the above bistable state.

Single-button on/off circuit

Fig. 4.6 shows how the *SCR* bistable can be converted for single button operation, so that one push of the switch turns the lamp on and the following push turns it off again. In this case, *SCR*2 has a large anode load, so its on current is lower than its minimum hold-on requirement; *SCR*2 is thus unable to latch on.

Assume that both *SCR*s are off; both anodes are near +ve rail voltage so zero charge is on C_1. When S_1 is operated, *SCR*1 and LP_1 are driven on via a brief +ve pulse from C_3, and *SCR*2 is momentarily driven on via a pulse from C_2. At the end of this brief pulse, *SCR*2 turns off again through lack of holding current, but *SCR*1 stays on. C_1 then charges via R_1, and *SCR*2 anode goes to +ve rail potential. The next time S_1 is operated, +ve pulses are again fed to both *SCR*s, but that on *SCR*1 gate has no effect, since *SCR*1 is already on. *SCR*2, on

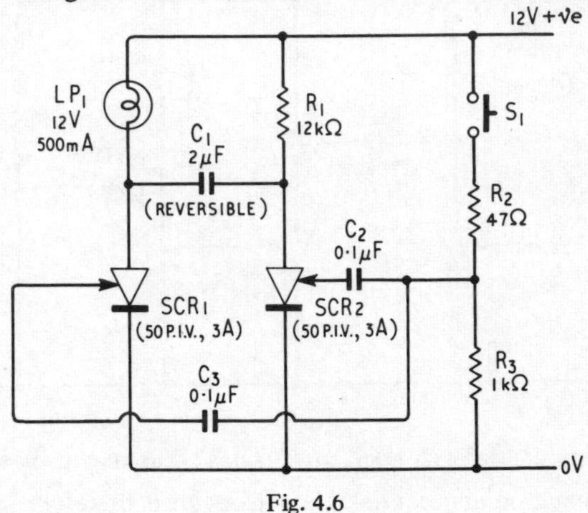

Fig. 4.6

Single-button on/off circuit

the other hand, is briefly driven on, and thus applies a reverse voltage to *SCR*1 via C_1, so *SCR*1 and LP_1 turn off. At the end of this pulse, *SCR*2 again turns off through lack of hold-on current, and the circuit is ready for the next operation of S_1.

The circuit changes state each time a +ve pulse is applied via S_1. Note therefore, that operation may become erratic if a noisy push-button is used. The possibility of erratic operation can be overcome by applying the trigger pulses via a one-shot transistor multivibrator.

Repetitive switching circuits

The circuit of Fig. 4.6 can be made to operate as a free-running or repetitive switch by feeding it with the trigger pulses from a u.j. pulse generator. Fig. 4.7 shows a practical version of a lamp flasher using this principle. This circuit gives equal on and off times of the lamp, and the repetition rate can be varied between about 25 and 150 flashes/min via R_5

A different type of flasher, giving independently variable on and off times, is shown in Fig. 4.8; the on and off times can be varied from

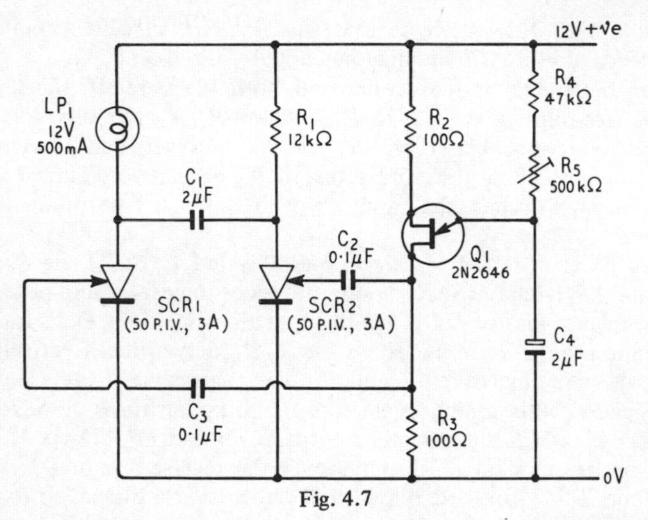

Fig. 4.7

Repetitive switching circuit, giving 25–150 flashes/min

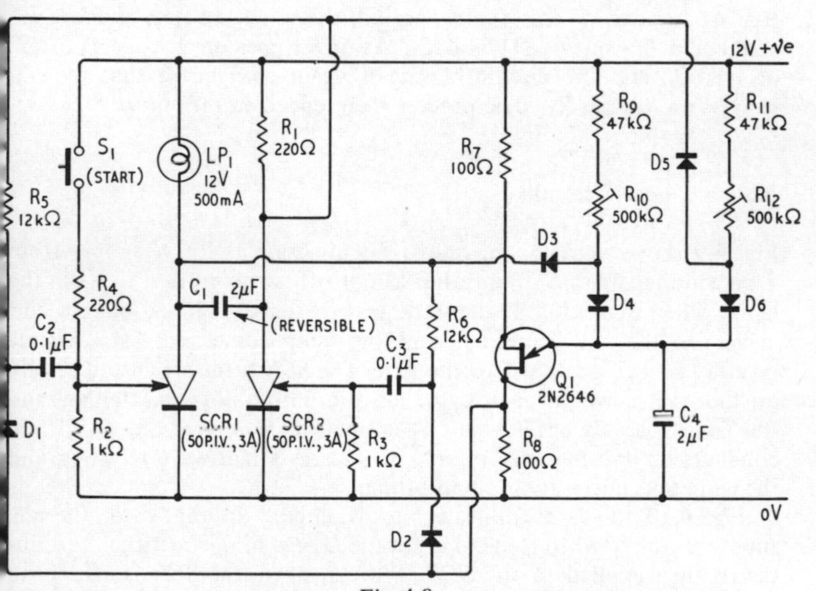

Fig. 4.8

Repetitive switching circuit, giving independently variable on/off times of 0.2–1.2 sec. D1–D6 are general purpose silicon diodes

approx. 0.2 to 1.2 sec. Note that this is a true bistable circuit, the anode loads of both *SCR*s being low enough for self-latching.

When the supply is first connected, both *SCR*s are off, and the u.j. timer is free-running via the R_9-R_{10}-*D*4 and R_{11}-R_{12}-*D*6 networks; *D*1 and *D*2 are reverse biased via R_5 and R_6, however, and prevent the trigger pulses reaching the *SCR* gates, so the u.j. has no practical effect at this stage. To start the circuit working, S_1 must be momentarily operated.

When S_1 is operated, a trigger pulse is fed to *SCR*1 via R_4, and *SCR*1 and LP_1 latch on. *SCR*1 anode thus goes to near-ground potential, and the reverse bias of *D*2 is removed; at the same time, *D*3 is forward biased and *D*4 is reverse biased so the R_9-R_{10} network is effectively cut out of the u.j. timer circuit, and the u.j. charges via R_{11}-R_{12} and *D*6. At the end of this timing cycle, the u.j. fires and turns on *SCR*2 via *D*2 and C_3; as *SCR*2 turns on, it turns *SCR*1 and LP_1 off via C_1. This puts a reverse bias on *D*2 but removes the reverse bias of *D*1; at the same time, *D*5 is forward biased and *D*6 is reverse biased, so R_{11} and R_{12} are effectively cut out of the u.j. timer circuit and *D*3 is reverse biased and *D*4 is forward biased, so the u.j. now charges via R_9 and R_{10}. At the end of this timing period, the u.j. again fires and triggers *SCR*1 and LP_1 on via *D*1 and C_2. As *SCR*1 goes on, it triggers *SCR*2 off via C_1, and the circuit biasing is again changed so that the u.j. charges via R_{11} and R_{12}. The process then repeats *ad infinitum.*

Basic a.c. on/off circuits

Fig. 4.9 shows a basic a.c. on/off circuit using a 12.6 V supply from a transformer. With S_1 open, the *SCR* is off, so no current flows in the lamp. When S_1 is closed, the *SCR* gate is forward biased on +ve half cycles, so the *SCR* conducts and the lamp comes on. *D*1 prevents reverse bias being applied to the gate. The *SCR* turns off automatically on the –ve halves of each cycle, so the unit is not self-latching, and the lamp goes off again when S_1 is opened. Note that the *SCR* only conducts on +ve half cycles, and so acts as a half-wave rectifier, and the lamp thus burns at only half brilliance.

Fig. 4.10 shows a full-wave on/off circuit. In this case, the a.c. supply is converted to rough d.c. via the *D*1-*D*4 bridge rectifier, and this d.c. is then applied to the *SCR*. With S_1 open, the *SCR* is off, so no current flows through the bridge via LP_1. When S_1 is closed, the *SCR* is biased on, so current flows through LP_1 via the bridge and *SCR*. The *SCR* voltage falls to zero once on every half cycle, so the circuit is not

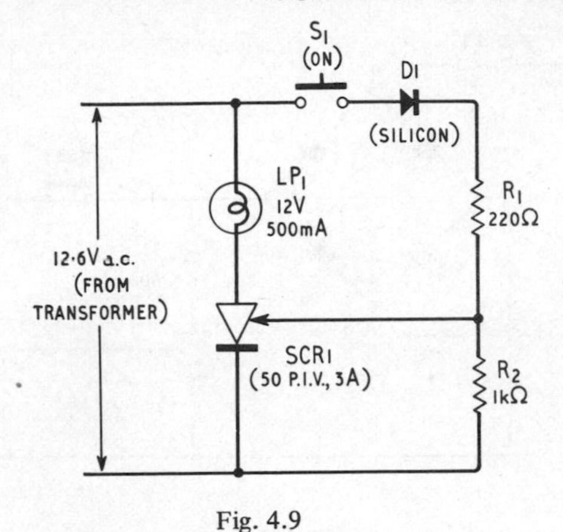

Fig. 4.9
Basic a.c. on/off circuit (Half-wave)

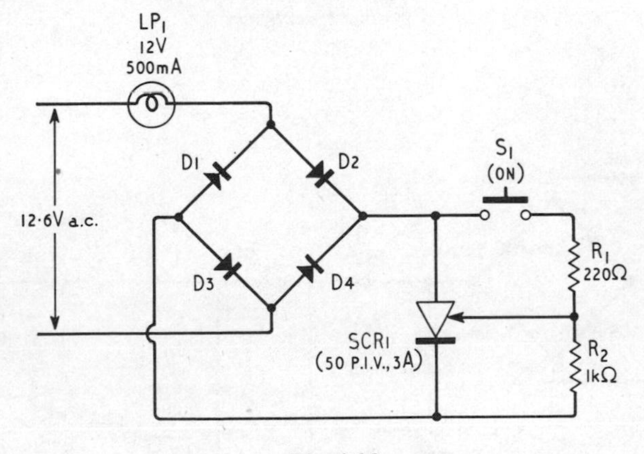

Fig. 4.10
Full-wave on/off circuit, controlling an a.c. load. D1–D4 *are 50 p.i.v., 3 A silicon rectifiers*

self-latching. Note that, in this circuit, LP_1 is on the a.c. side of the bridge, while the *SCR* is on the d.c. side, so the design is in fact used to control an a.c. load.

The circuit of Fig. 4.11 is similar to that of Fig. 4.10, but in this case the lamp load is wired in series with the *SCR* anode, so this design is used to control a d.c. load. The circuit is not self-latching.

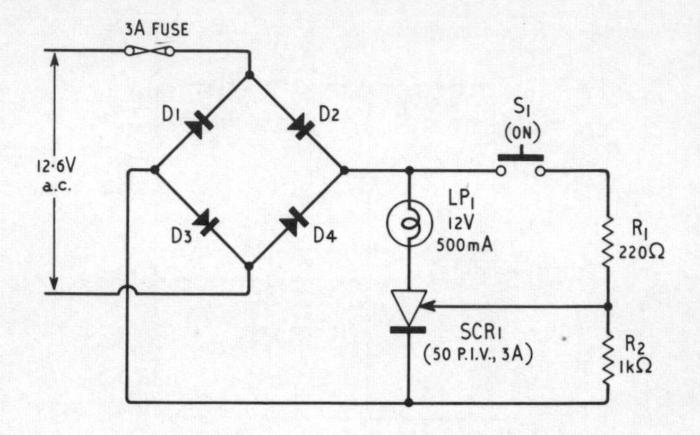

Fig. 4.11

Alternative full-wave on/off circuit, controlling a d.c. load. D1–D4 *are 50 p.i.v., 3 A silicon rectifiers*

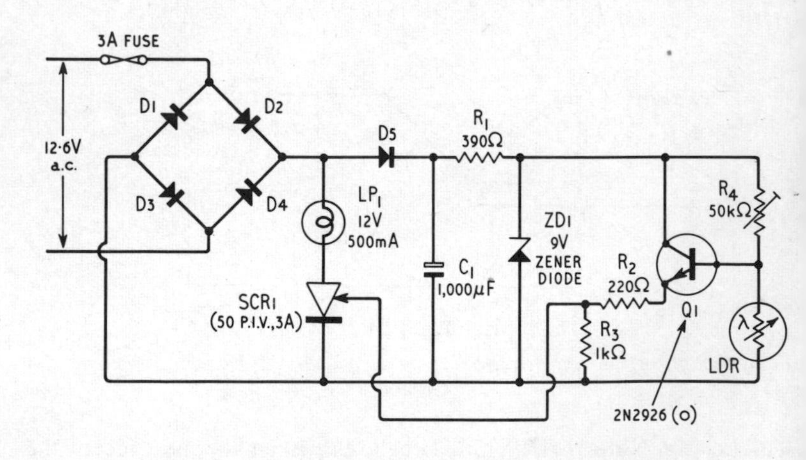

Fig. 4.12

Light-operated switch (non-latching) D1–D4 *are 50 p.i.v., 3 A silicon rectifiers,* D5 *is a general purpose silicon diode, and* LDR *is any cadmium sulphide photocell with a face diameter greater than 0.25 in*

Light-operated SCR switches

Fig. 4.12 shows how Fig. 4.11 can be converted for use as a light-operated switch. Here, the *SCR* and lamp are fed with rough d.c. from the bridge rectifier, and this is then smoothed by C_1 but prevented from reaching the *SCR* by *D*5. The smoothed d.c. is then stabilised at 9 V at 20 mA via R_1 and *ZD*1, and is used to power the *Q*1 transistor circuitry. *Q*1 is wired as an emitter follower, with base-bias provided via potential divider R_4 and *LDR*. Under bright conditions, the *LDR* resistance is low, so the voltage on *Q*1 emitter is not sufficient to trigger the *SCR*, and LP_1 is off. Under dark conditions, the *LDR* resistance is high, so the voltage on *Q*1 emitter is sufficient to trigger the *SCR*, and LP_1 comes on. Since the *SCR* is fed with rough d.c. the circuit is not self-latching, and the lamp turns off when the gate bias is removed.

Fig. 4.13 shows how Fig. 4.12 can be modified for self-latching operation. Here, when the *SCR* is on, it passes the rough current of

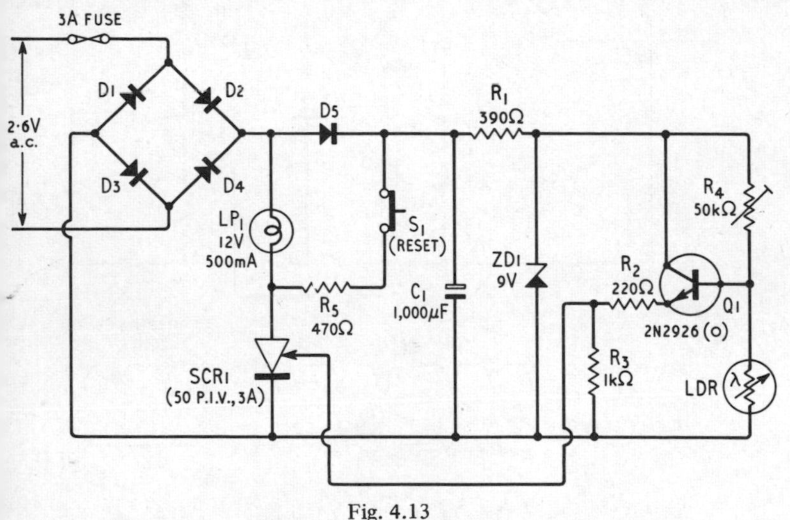

Fig. 4.13

Modification of Fig. 4.12 giving self-latching operation D1–D4 *are 50 p.i.v., 3 A silicon rectifiers*

the lamp plus a low but smoothed 'standby' current from R_5. The standby current, however, is greater than the *SCR*s minimum holding current, so, once the *SCR* has been driven on, the gate loses control, and the lamp stays on even though the gate bias is removed. The *SCR* can only be turned off by removing the gate bias and disconnecting the holding current by operating 'reset' button S_1.

The circuits of Figs. 4.12 and 4.13 can be modified for operation by sound, heat, etc., by simply replacing $Q1$ with alternative detector circuitry.

Variable-power circuits

Fig. 4.14a shows how the *SCR* can be used, in conjunction with a u.j. pulse generator, as a variable power unit feeding a d.c. load. The circuit

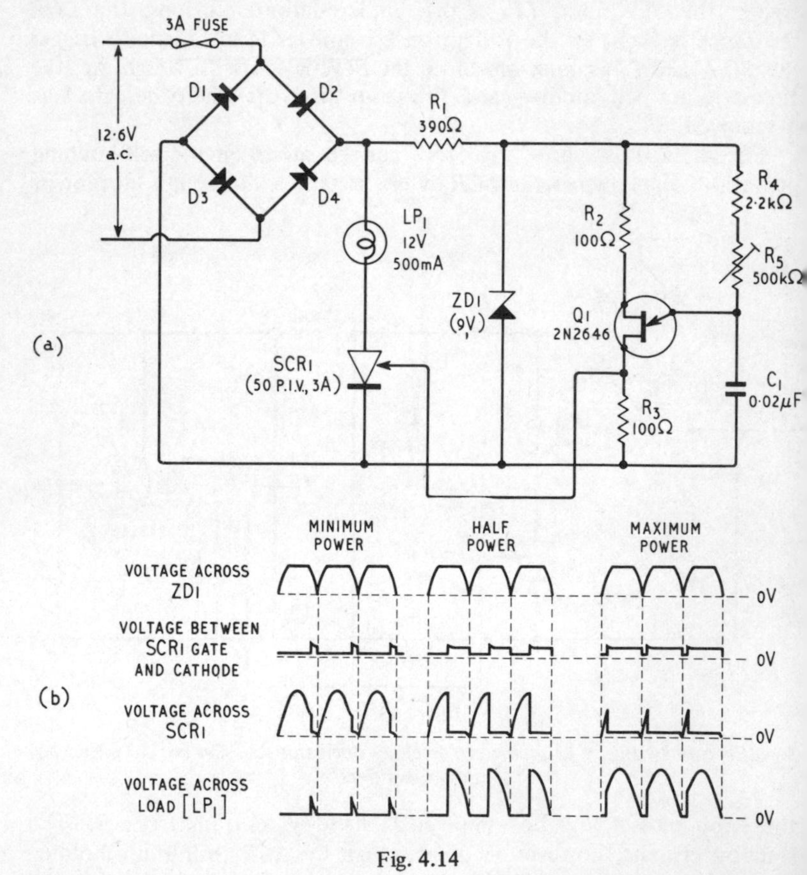

Fig. 4.14

(a) Variable-power unit, feeding a d.c. load. D1–D4 *are 50 p.i.v., 3 A silicon rectifiers. (b) Wave-forms of Fig. 4.13a under alternative operating conditions*

waveforms are shown in Fig. 4.14b. Here, the voltage across *ZD*1, and thus across the u.j. circuit, is rough d.c. clipped at 9 V, so the power to the generator is automatically connected and disconnected in sympathy with the power line frequency. At the start of each new half cycle, the u.j. circuit starts a timing cycle, and, after a delay determined

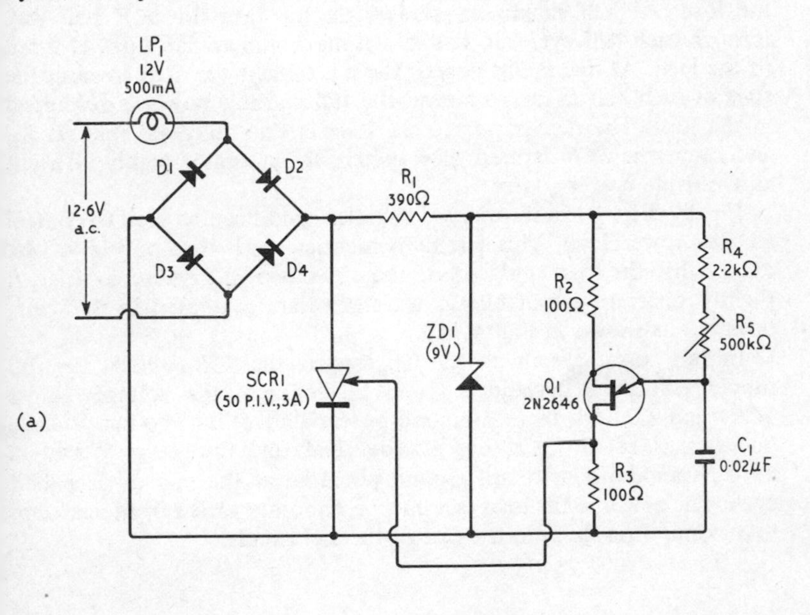

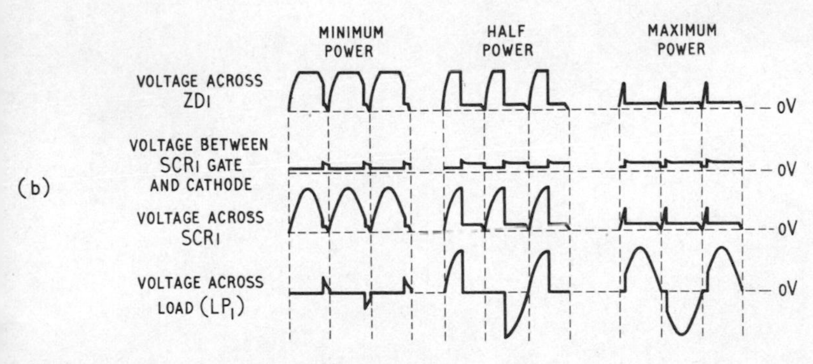

Fig. 4 15

(a) Variable-power unit, feeding an a.c. load. D1–D4 *are 50 p.i.v., 50 silicon rectifiers. (b) Wave-forms of Fig. 4.1a, under alternative operating conditions*

by R_5, generates a +ve pulse and fires *SCR*1. Thus, the u.j. gives delayed and variable firing of the *SCR*.

When the unit is set for minimum output power (in LP_1), the u.j. gives maximum delay, so the *SCR* fires towards the very end of each half cycle, so only a small part of the total available power is fed to the load. At half maximum power, the u.j. fires the *SCR* half way through each half cycle, so half of the maximum available power is fed to the load. At maximum power, the u.j. triggers the *SCR* towards the start of each half cycle, so almost the full available power is developed in the load. The d.c. power to the load is thus fully variable via R_5, and, since the *SCR* is used as a switch, the system is highly efficient as a variable power source.

Finally, Fig. 4.15a shows how a similar circuit can be used to control an a.c. power load. This circuit is identical with that of Fig. 4.14a, except that the load is placed on the a.c. side of the bridge rectifier. A slightly different set of circuit waveforms are generated in this case, however, as shown in Fig. 4.15b.

In this case, as soon as the u.j. triggers the *SCR*, almost the full supply voltage is developed across the load, so the voltages across *SCR*1 and *ZD*1 fall to near-ground potential. This is of no importance, however, since the *SCR* has already fired, and thus stays locked-on until its anode falls to full ground potential at the end of each half cycle. The power to the load can thus be smoothly varied from near-zero to maximum via R_5, as in the case of the d.c. circuit.

CHAPTER 5

30 INTEGRATED CIRCUIT PROJECTS

The actual semiconductor 'heart' of a transistor is physically very small. So small, in fact, that it can be clearly seen only with the aid of a microscope. The physical size of the complete transistor, however, is dictated by the practical need of human operators comfortably to handle the device, and to meet this need the 'heart' is usually shrouded in a relatively massive case, and is connected to equally massive external leads. Thus, although the final transistor is quite small by most standards, the relative size of the 'heart' to the case compares, by analogy, to that of an orange to a household garbage can. There is, in fact, enough room in the average sized transistor case to hold hundreds of semiconductor 'hearts'.

A similar thing is true of resistors. Most of the volume of a conventional resistor is taken up by a 'body' or former, and on the outside of this is a thin film of carbon or similar material which forms the true resistance. The volume of resistance material is very small relative to that of the body.

Now, it follows from the above that, if the need to handle individual transistors can be eliminated, and the need to have an individual body for each resistor can be eliminated as well, it should be possible to produce a complete circuit, with many 'transistors' and 'resistors', in a single case the size of one conventional transistor. Only a few external connections need be made to such a circuit; in a linear amplifier, for example, the only connections needed are those of input, output, and power supplies. Thus, the idea seems feasible, and in the past decade the technology has indeed been developed to put the idea into practice. The devices embodying the idea are known as 'integrated circuits', or i.c.s.

An interesting point about i.c.s is that, because of the manufacturing techniques used, the actual tooling costs of producing an i.c. using, say, fifty 'transistors', is no greater than that of an i.c. using only two 'transistors'. The actual cost of the i.c. to the consumer, therefore, is dictated almost entirely by the sales volume of the individual device type. If the i.c. is that of a highly specialised circuit, in which total sales are only likely to reach a few thousand, the unit-cost must be high, since the manufacturer must impose a high surcharge per unit in order to recover the capital outlay of tooling up, etc., and still make

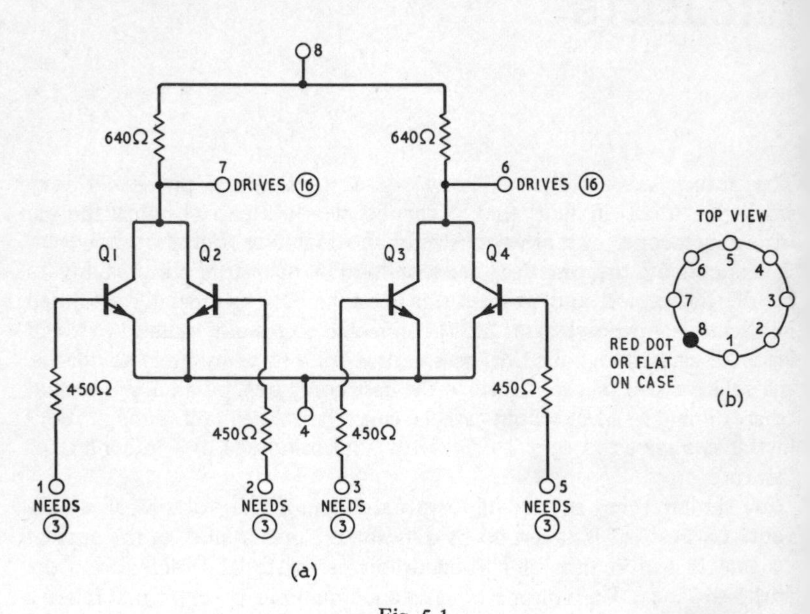

Fig. 5.1

(a) Internal circuit of μL914. (b) Lead connections of μL914

a profit on the limited sales. If, on the other hand, the i.c. is that of a popular and versatile circuit, in which sales may reach the million mark, unit costs can be very low, since the same capital outlay is involved and can be recovered with a proportionally low surcharge per unit. Many different types of i.c. are thus in production, and cover a very wide price range.

The least expensive of all i.c.s are the digital or logic types, which are widely used in computers and consequently have a high sales volume.

Some of these units are quite versatile, and can be used in a number of applications of interest to the amateur experimenter. Of particular interest in this respect is the μL914 'dual gate' integrated circuit by Fairchilds. This unit is one of their so-called 'micro-logic', or μL line, and costs little more than a single conventional transistor. Fig. 5.1a shows the internal circuit of the unit, which incorporates four silicon planar 2N708 transistors and six resistors in a single TO5-sized eight lead epoxy resin case. The 2N708 transistors have an f_T, or gain-bandwidth product, of 450 MHz. Fig. 5.1b shows the lead connections of this i.c., which is used as the basis of all 30 circuits described in this chapter.

Using integrated circuits

There are just two basic rules to remember when designing i.c. projects. First, do *not* think of the i.c. as a special semiconductor device needing new design techniques all of its own. Second, *do* think of the i.c. as a bunch of quite normal components, all bonded together in a single block, and requiring perfectly normal transistor circuit design techniques.

With the second rule in mind, take a look at the internal circuit of the μL914 shown in Fig. 5.1a. Although this unit is specifically intended for use as a gate or logic circuit, it can in fact be made to work in many other applications. Unwanted transistors can be effectively cut out of circuit by shorting them base to emitter, so this i.c. can be made to work as a simple common emitter amplifier, using *Q*1 only, by shorting pins 2, 3, 4 and 5 together (thus cutting *Q*2, *Q*3 and *Q*4 out of circuit) and then wiring a single base-bias resistor between pins 1 and 7. Alternatively, it can be made to work as a dual common emitter, using *Q*1 and *Q*4, by shorting *Q*2 and *Q*3 out of circuit and applying bias to *Q*1 and *Q*4.

Again, by cutting *Q*3 and *Q*4 out of circuit and applying bias to *Q*1 and *Q*2, the i.c. can function as a 2-channel mixer, with output taken from pin 7. Also, using *Q*1 and *Q*4 only, it can be made to function as a differential amplifier or phase splitter, or, if cross-coupling is used between *Q*1 and *Q*4, it can be made to function as an astable, monostable, or bistable multivibrator.

The μL914 can be used in a variety of pulse and logic applications. An interesting feature of the entire 'micro-logic' line of i.c.s is that input and output drives do not have to be worked out in terms of impedance and current when connecting the i.c.s to one another. Instead, Fairchilds specify input needs and output drive as so many 'units' of drive, as shown circled in the diagrams.

Thus, when used as a 'dual gate', the μL914 needs 3 units of drive at each input and gives 16 units of drive at each output, so each output can drive as many as five following input stages. Thus, if banks of μL914s are to be interconnected, it is easy to calculate just how many inputs each output can drive directly, and normally complex circuit layout calculations can be simply made.

Having cleared up these points, we can now go on and look at 30 practical applications of the μL914.

Pulse inverter and gate projects

The μL914 can be used as a pulse inverter by cutting out all transistors except one, and then connecting the input to the base of this remaining transistor and taking the output from its collector. In the absence of a pulse (input grounded) this transistor is cut off and its collector output is fully +ve; when a positive pulse input is applied, the transistor is driven to saturation and its collector output falls to ground potential.

Fig. 5.2a shows how to connect the i.c. as a pulse inverter using *Q*1 only, and Fig. 5.2b shows an alternative connection using *Q*4. Fig. 5.3 shows the connections for a dual pulse inverter, using both *Q*1 and *Q*4.

Incidentally, the 450 Ω resistors that are connected between the base junctions of the i.c. and the external connecting leads prevent excessive base currents flowing when the base leads are shorted to the +ve supply line, and this feature makes the i.c. almost indestructible, providing that supply potentials greater than about 6 V are not used.

Fig. 5.4 shows how to connect the i.c. as a pulse disabling gate. Here, *Q*3 and *Q*4 are cut out of circuit, and the output is taken from pin 7. When a positive input pulse train is applied to pin 1, and pin 2 is grounded, an inverted pulse output is available via *Q*1, as in the case of Fig. 5.2a. When, on the other hand, pin 2 is made +ve by a gate signal, *Q*2 is driven to saturation and the output is locked to ground potential, so no output is available from the pin 1 input; the circuit is thus disabled.

Fig. 5.6 shows how to connect the i.c. as a dual pulse disabling gate, using all four transistors.

The output of the Fig. 5.4 circuit is inverted relative to the pin 1 input. Fig. 5.5 shows how to connect the i.c. as a non-inverting pulse disabling gate, if required.

An alternative type of gate circuit is the pulse enabling gate shown in Fig. 5.7. Here, with an input signal applied to pin 1, there is no

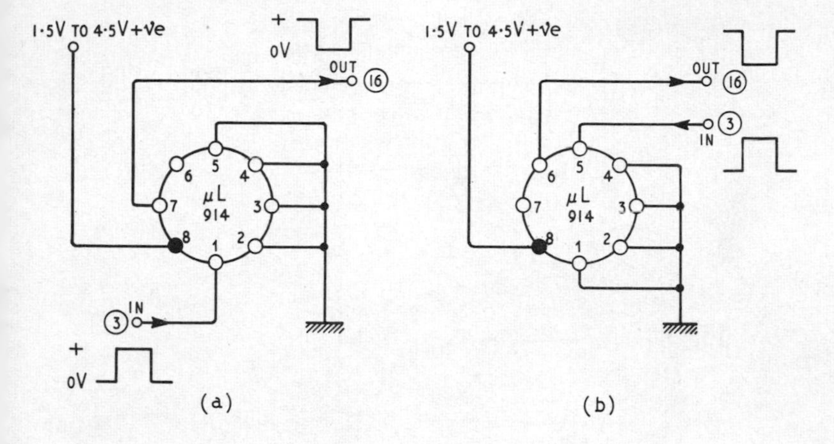

Fig. 5.2

(a) Pulse inverter. (b) Alternative pulse inverter

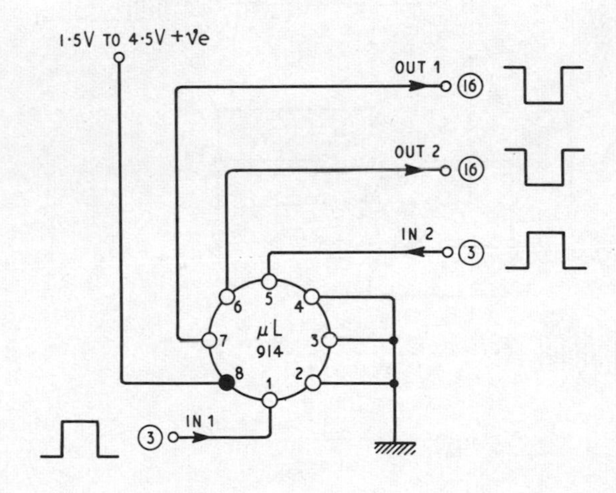

Fig. 5.3

Dual pulse inverter

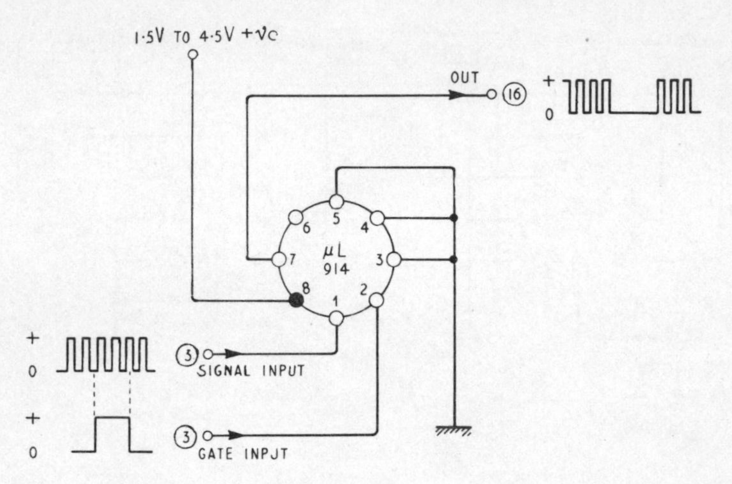

Fig. 5.4

Pulse disabling gate (inverting)

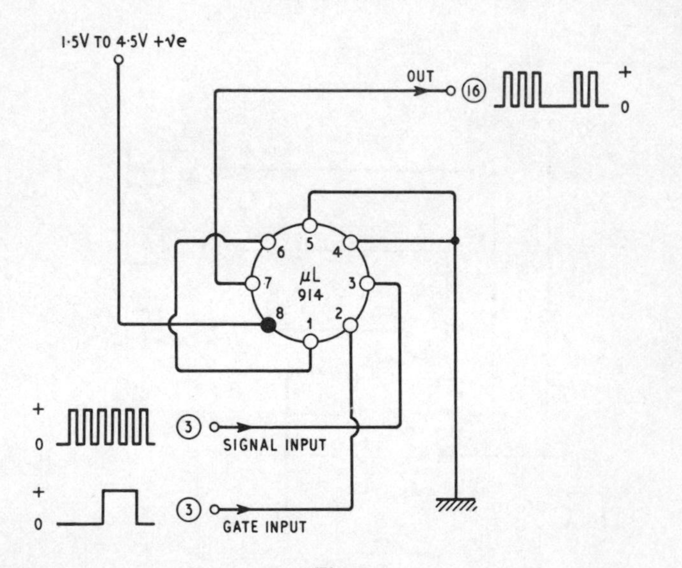

Fig. 5.5

Pulse disabling gate (non-inverting)

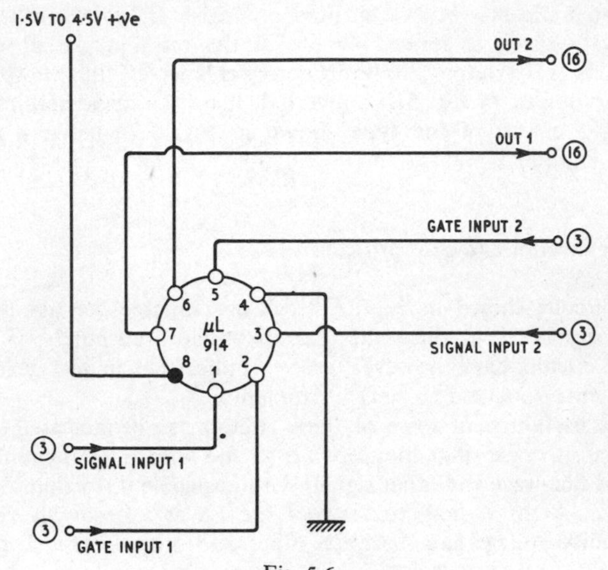

Fig. 5.6

Dual pulse disabling gate (inverting)

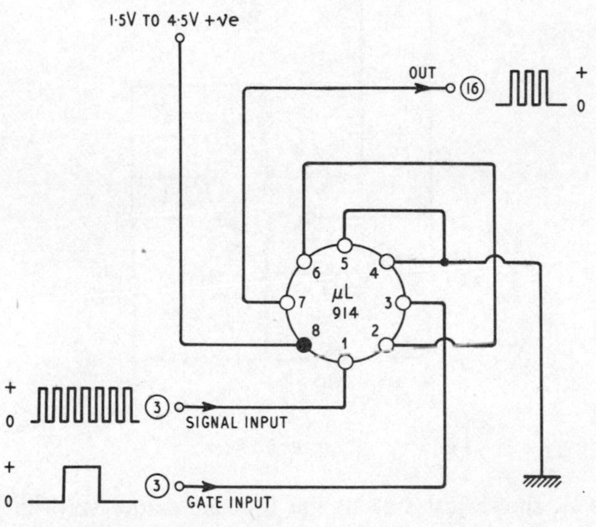

Fig. 5.7

Pulse enabling gate (inverting)

output if pin 3 is held at ground potential. If, on the other hand, a +ve gate signal is applied to pin 3, the pin 1 signal does become available at the output; the +ve gate signal 'enables' the gate to open.

The output of Fig. 5.7 is inverted. It can be made non-inverted by wiring a circuit of the type shown in Fig. 5.2a between the input signal and pin 1.

Linear inverter and gate projects

The circuits shown in Figs. 5.2–5.7 are suitable for use with pulse input signals only, since the transistors are used purely as switches. These circuits have, however, many applications in test gear circuits, tone generators, and musical instruments.

The usefulness of some of these circuits can be increased by biasing the transistors so that they operate in the linear mode, enabling them to pass sine-wave and other signals with negligible distortion.

Fig. 5.8 shows how to connect the i.c. as a linear inverter, giving near-unity voltage gain between input and output. This diagram, like

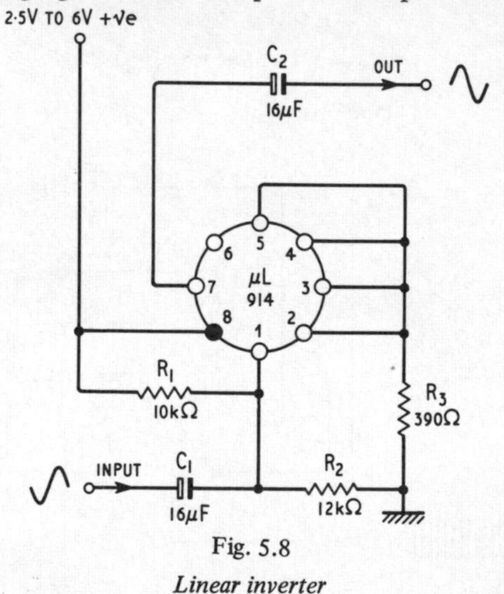

Fig. 5.8
Linear inverter

Fig. 5.2a, shows just one of the four alternative ways in which the inverter can be connected. If exact unity voltage gain is required from this circuit, R_3 should be replaced with a 270 Ω resistor and 250 Ω potentiometer in series, and the potentiometer should be adjusted to give the correct gain.

Fig. 5.9 shows how to wire the circuit as a linear disabling gate.

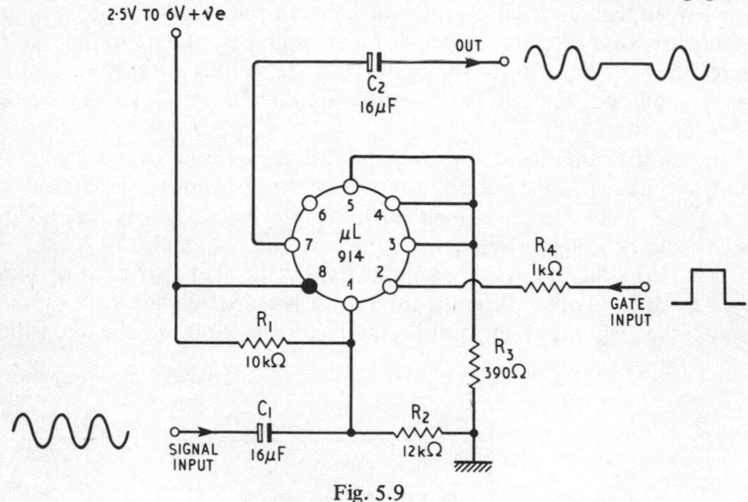

Fig. 5.9

Linear disabling gate

This can be converted to a linear enabling gate by wiring a Fig. 5.2a type inverter between the gate-pulse input and R_4.

Figs. 5.8 and 5.9 have useful applications in audio and sound-distribution systems, and in oscilloscope trace doublers and frequency measuring gear.

Logic circuits

The µL914 can perform all the basic functions used in computer logic. Before looking at the different i.c. connections, however, a brief explanation must be given of the meanings of the different logic terms.

In logic work, inputs and outputs are either fully on (+ve or –ve) or off (zero or grounded). The state of the output depends on the way the different inputs are connected. If we have a two-input circuit, with inputs *A* and *B*, and an output becomes available when either *A or B* is connected, we call that an OR logic circuit. If, on the other hand, the output appears only when both *A and B* are connected, we call it an AND logic circuit.

Note that no attempt has been made to define an input or output in the above explanation. That is because the subject starts to get pretty complicated when we start specifying these conditions. To simplify further explanations, however, we can confine explanations to npn transistor logic circuits, in which inputs and outputs are either +ve or zero. In this case, if both the inputs and the output are specified in the same condition, the two basic circuits are known simply as AND or OR logic circuits.

If, on the other hand, different conditions are used to specify inputs and output, i.e., inputs zero and output +ve, or input +ve and output zero, the logic circuit names must be prefixed with an 'N', giving NAND and NOR logic circuits in addition to the AND and OR types.

The situation is further complicated by the fact that each of these four circuits has two different variations. An AND circuit, for example, may give a +ve output when both inputs are +ve, or it may be one which

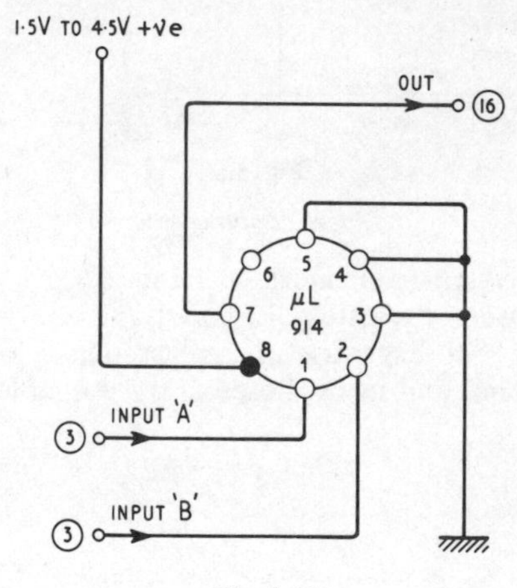

Fig. 5.10

NOR/NAND Logic circuit. Output is grounded if A *or* B *is +ve: output is +ve if* A *and* B *are grounded*

gives a zero output when both inputs are zero. Thus, although there are only four basic logic types, there are eight variations of logic.

A final complication is that each practical circuit can be known by either of two names. If, for example, an AND circuit gives a +ve output only when *A* and *B* are +ve, it follows that the output must be zero when *A* or *B* is zero, so the circuit can also be used to perform OR logic functions.

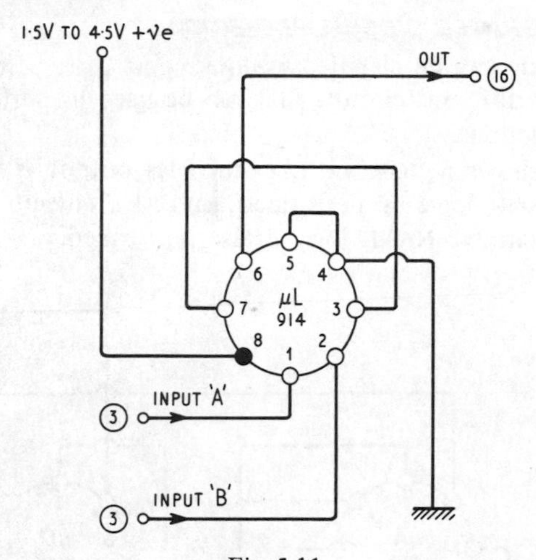

Fig. 5.11

OR/AND Logic circuit. Output is +ve if A *or* B *is +ve: output grounded if* A *and* B *are grounded*

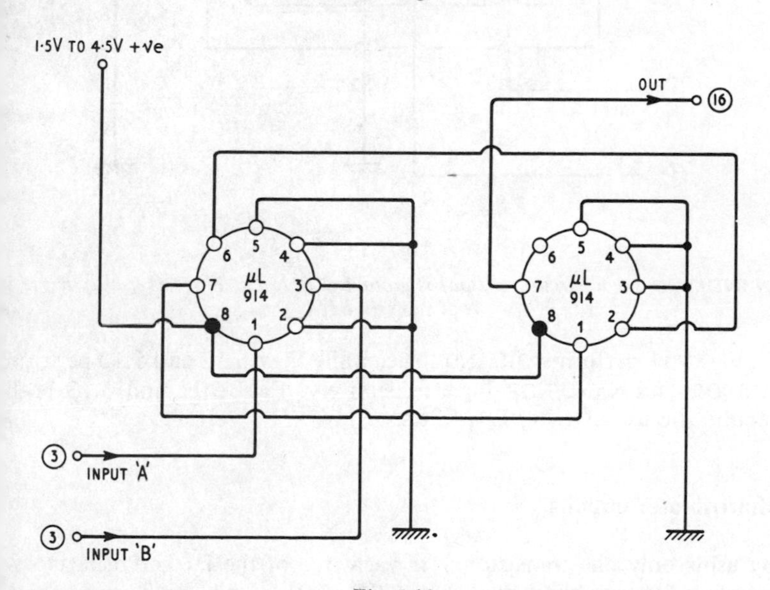

Fig. 5.12

AND/OR Logic circuit. Output +ve if A *or* B *are +ve: output is grounded if* A *or* B *is grounded*

This situation can be clarified by looking at Figs. 5.10–5.13, which show the four different circuits that can be used to perform the eight basic logic functions.

Fig. 5.10 shows a NOR/NAND circuit. Its output is zero if *A* or *B* are +ve, so NOR logic is performed, and the output is +ve only if *A* and *B* are zero, so NAND logic is also performed.

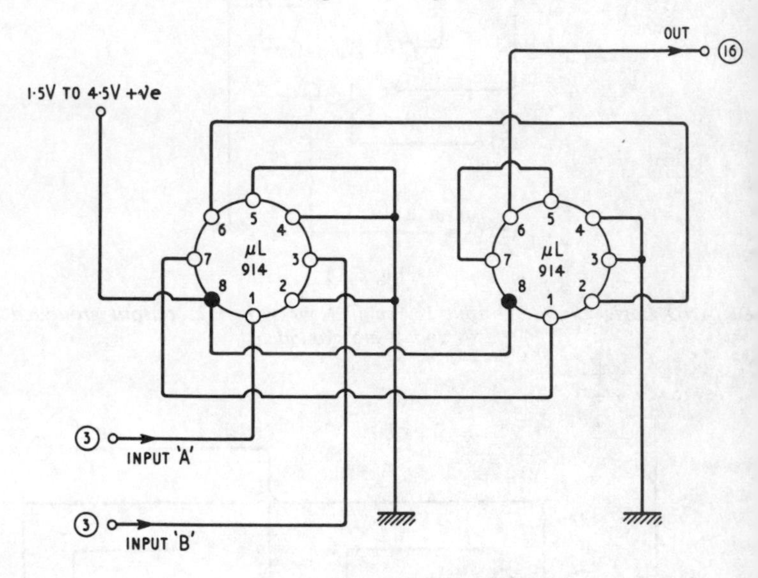

Fig. 5.13

NAND/NOR Logic circuit. Output is grounded if A *and* B *are +ve: output +ve if* A *or* B *is grounded*

Fig. 5.11 performs OR/AND logic, while Figs. 5.12 and 5.13 perform AND/OR and NAND/NOR logic respectively. Figs. 5.12 and 5.13 both require the use of two μL914 i.c.s.

Multivibrator circuits

By using only one transistor from each side of the i.c. and using cross-coupling between these two stages, the μL914 can be made to perform as any one of the three basic multivibrator types, i.e., it can work as a free-running or astable multivibrator, as a monostable multivibrator or triggered pulse generator, or as a bistable multivibrator or memory unit.

Fig. 5.14 shows how to connect the i.c. as a 1 kHz astable multivibrator or waveform generator. C_1 and C_2 are the cross-coupling capacitors, and R_1 and R_2 are the capacitor discharge resistors. The operating frequency of the multivibrator can be increased by reducing the values of C_1 and C_2, or reduced by increasing the values of these components. If the values are reduced by a factor of 10 (to 0.01 μF), the frequency

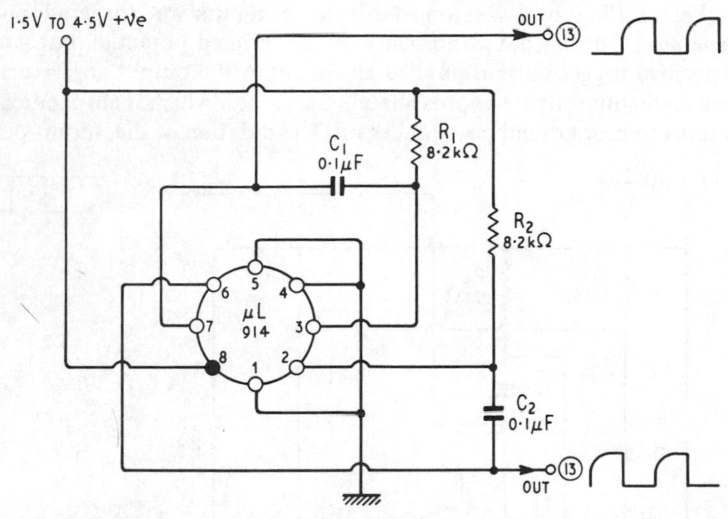

Fig. 5.14
1 kHz astable multivibrator

increases by a decade (to 10 kHz). The output waveform, which is approximately rectangular but has a slightly rounded leading edge, can be taken from either pin 7 or 6.

The rounding of the leading edge of the output waveform, which occurs to some degree in all simple astable multivibrator circuits, is caused by the fact that, at the moment that the circuit switches from one state to the other, the collector of the 'off' transistor is momentarily pulled down towards the base of the 'on' transistor via the cross-coupling capacitor. This rounding can be eliminated by disconnecting the output from the capacitor at the moment that the circuit changes state, and this can be achieved by using the connections shown in Fig. 5.15.

Fig. 5.15 generates a near-perfect square wave, with no rounding of the leading edge. Using the component values shown, the circuit operates at a frequency of about 1 kHz. The frequency can be altered by changing the values of C_1 and C_2, as in the case of Fig. 5.14.

Fig. 5.16 shows how to connect the i.c. as a bistable multivibrator or memory unit. If a +ve pulse is applied to pin 2, the output at pin 6 goes +ve and at pin 7 goes to zero, and the outputs remain in this state until a +ve reset pulse is applied to pin 3, at which time pin 6 goes to zero and pin 7 goes +ve, the circuit remaining in this state until another +ve pulse is applied to pin 2.

Fig. 5.17 shows a monostable multivibrator or triggered pulse generator. The output is normally at near-ground potential, but when a positive trigger pulse is applied to the input the output goes +ve and remains in this state for approximately 5 sec, after which it automatically returns to near-ground potential again. The duration of the output pulse

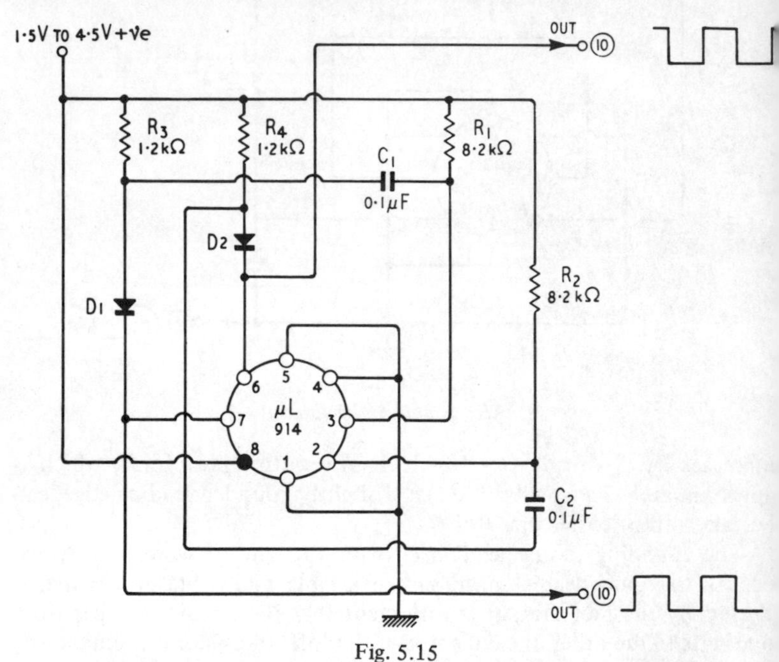

Fig. 5.15
1 kHz square-wave generator. D_1 *and* D_2 *are any germanium diodes*

is controlled by C_1, and can be varied from a few micro-seconds to several seconds by selection of this component value.

The monostable multivibrator can be used as a 'noiseless' push button by wiring R_3 and S_1 (shown dotted) in place as indicated. A normal push button, of course, generates a good deal of noise, due to contact bounce, which may be harmful in some applications.

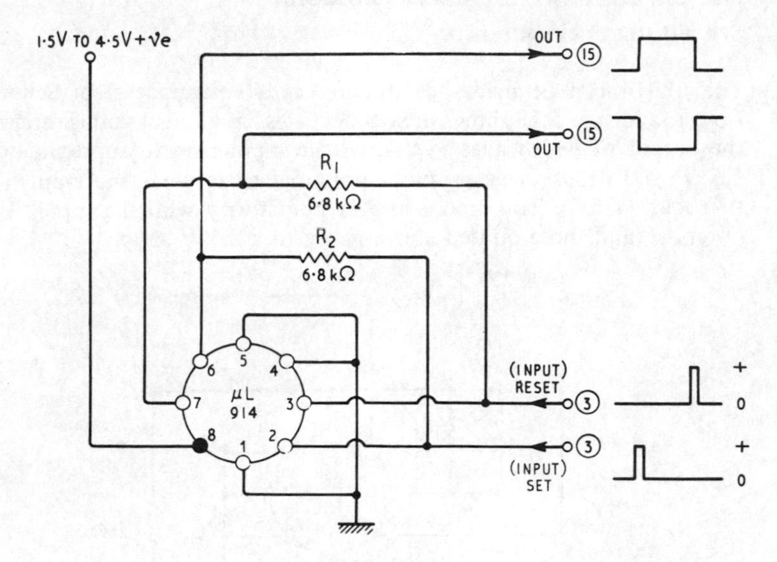

Fig. 5.16

Bistable multivibrator or memory unit

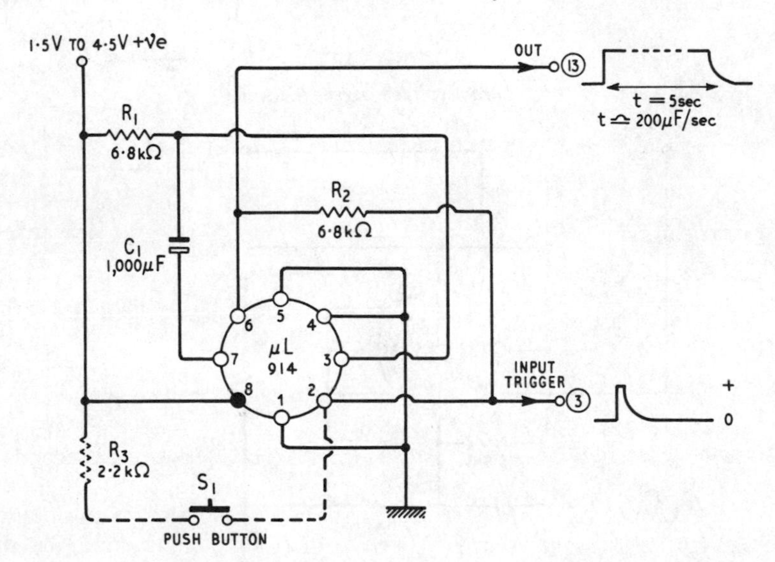

Fig. 5.17

Monostable multivibrator

Schmitt trigger circuits

The μL914 can be made to function as a voltage level or Schmitt trigger, and Fig. 5.18 shows the connections for a direct voltage trigger, the output of which goes +ve when the input rises to approximately 1.5 V, and drops to near-ground potential again when the input falls to about 1.25 V. The precise trigger points vary with the supply line potential, and those quoted above apply to a 3.5 V supply.

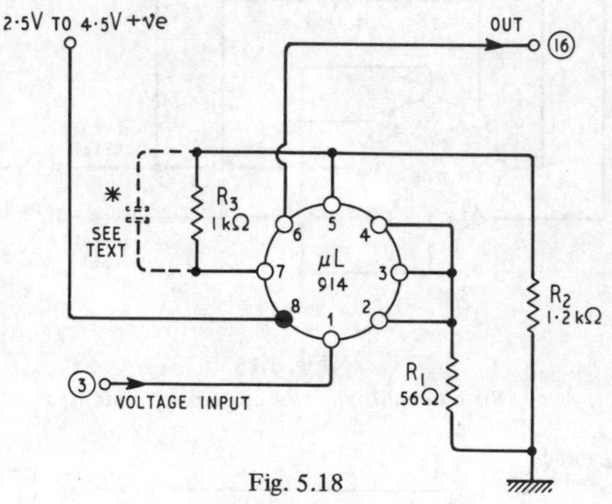

Fig. 5.18

Direct voltage trigger (Schmitt)

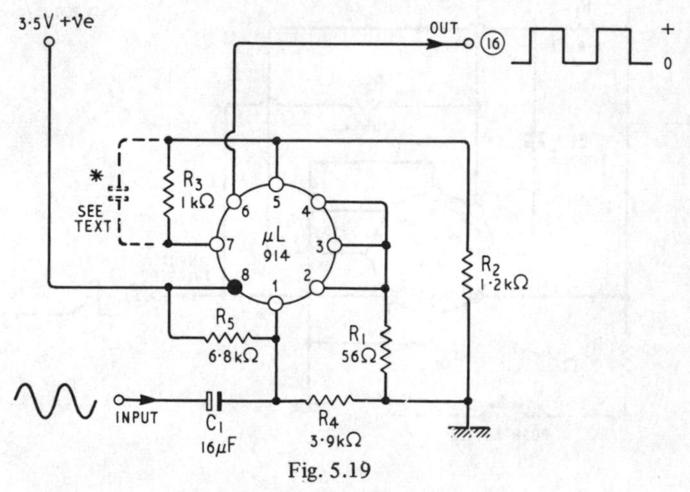

Fig. 5.19

Sine/square converter

The Schmitt trigger can be used as a sensitive sine/square converter by using the connections shown in Fig. 5.19. This circuit gives an output of approximately square form when a sine wave of amplitude greater than about 150 mV is applied to the input.

The two circuits shown in Figs. 5.18 and 5.19 can be operated at frequencies up to several hundred kiloHertz. At frequencies above a few tens of kiloHertz, the output waveform can be improved by connecting a small capacitor (shown dotted) across R_3. The value of this capacitor should be found by trial and error, to give the best output wave shape, but generally has a value of the order of 100 pF.

Linear phase splitters

Fig. 5.20 shows how to wire the μL914 as a simple phase splitter, giving near-unity voltage gain. Only one of the i.c. transistors is used, and is

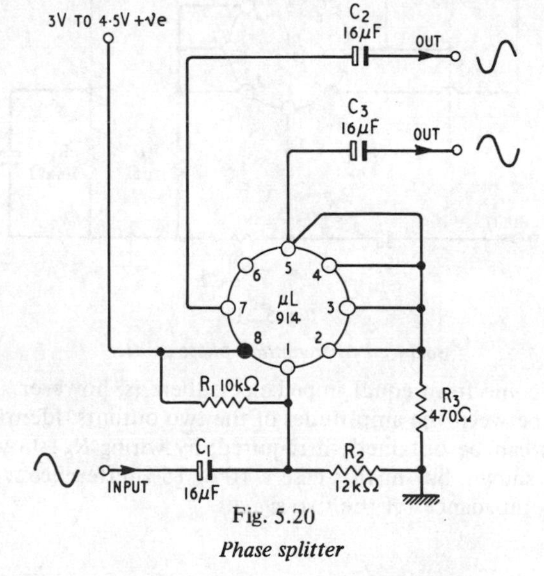

Fig. 5.20
Phase splitter

wired as a common emitter amplifier with emitter degeneration provided by R_3. The two output signals of the unit are roughly equal in amplitude and are in anti-phase, but are at different impedance levels.

If the two outputs are required to be exactly equal in amplitude, R_3 should be replaced by a 390 Ω resistor and 250 Ω potentiometer in series, the potentiometer then being adjusted for an exact output balance. Peak-to-peak input signals of about 1 V can be handled by this circuit when a 3 V supply is used.

Fig. 5.21 shows the connections for a balanced or differential phase splitter, using two of the transistors in the i.c. This circuit gives a voltage gain of about 8 times, and the two output signals are in anti-

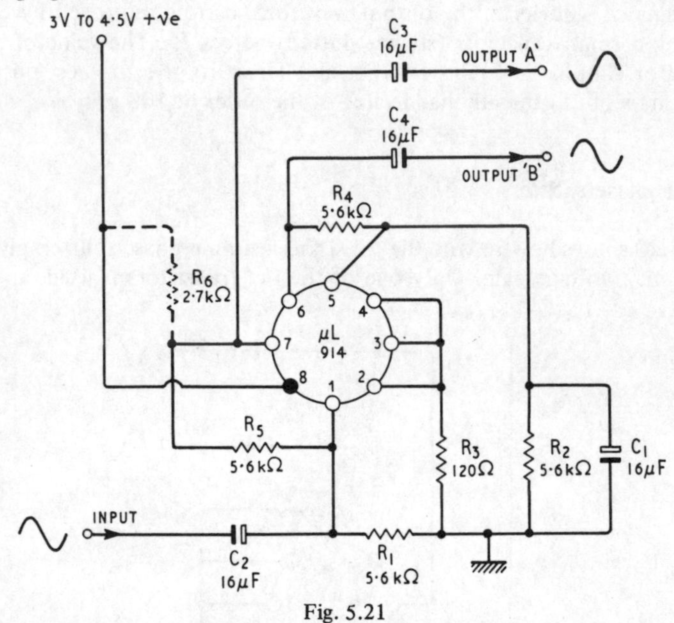

Fig. 5.21

Balanced (differential) phase splitter

phase and come from equal impedances; there is, however, 10 or 15% difference between the amplitudes of the two outputs. Identical output amplitudes can be obtained, if required, by wiring R_6 (shown dotted) in place as shown, but in this case a 10 or 15% difference will occur in the output impedances of the two signals.

Waveform generators

When the μL914 is operated as a differential amplifier, as in the case of Fig. 5.21, one of the two outputs is in phase with, but of greater amplitude than, the input signal. Consequently, positive feedback occurs if this output is coupled back to the input, and in this case

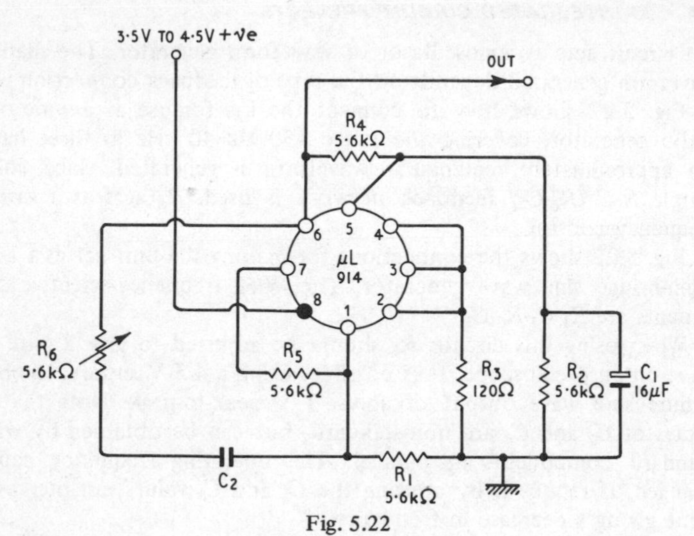

Fig. 5.22

Wide range audio generator, if C_2 = 1 μF, f = 150 Hz–3 kHz; if C_2 = 0.1 μF, f = 3–15 kHz; if C_2 = 0.02 μF, f = 12–40 kHz

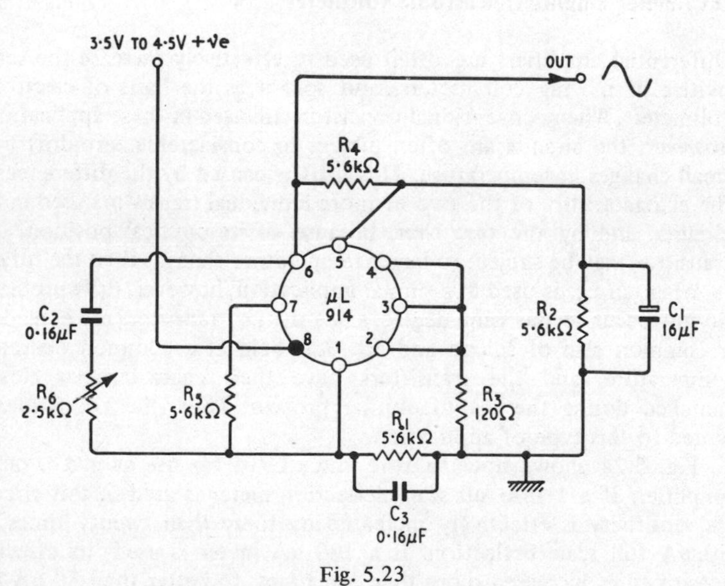

Fig. 5.23

1 kHz Wien bridge sine wave oscillator

the circuit acts as an oscillator or waveform generator. The shape of waveform generated depends on the type of feedback connection used.

Fig. 5.22 shows how to connect the i.c. for use as a wide range audio generator, covering the range 150 Hz–40 kHz in three bands. An approximately rectangular waveform is generated, since only a simple *R-C* $(R_6\text{-}C_2)$ feedback network is used. R_6 acts as a variable frequency control.

Fig. 5.23 shows the connections for making the unit act as a 1 kHz Wien-bridge sine wave generator. The Wien frequency-selective components are $R_1\text{-}C_3\text{-}R_6\text{-}C_2$.

When using this circuit, R_6 should be adjusted to give a pure sine wave on an oscilloscope. It is possible, using a 4.5 V supply, to obtain a pure sine wave output of about 1 V peak-to-peak. Note that the values of C_2 and C_3 are non-standard, but can be obtained by wiring standard components in parallel. The operating frequency can be changed, if required, by altering the C_2 and C_3 values, an increase in value giving a decrease in frequency.

D.C. meter amplifier/electronic voltmeter

Differential amplifiers are often used to effectively increase the sensitivities of moving coil meters, and so act as the basis of electronic voltmeters. When conventional transistors are used in these applications, however, the circuits are often subject to considerable zero-drift with small changes in temperature. This drift is caused by the differences in the characteristics of the two or more individual transistors used in the circuits, and by the fact that, because of its physical position, one transistor may be subject to larger temperature changes than the others.

When an i.c. is used in a similar application, however, drift problems do not occur to the same degree, since the i.c. transistors are etched on a common slab of silicon and are thus held at a common operating temperature, and the transistors have their characteristics closely matched during the manufacturing process. Thus, the i.c. is ideally suited to this type of application.

Fig. 5.24 shows how to wire the μL914 for use as a d.c. meter amplifier. If a 1 mA full scale deflection meter is used in this circuit its sensitivity is effectively increased by more than twenty times, to 50 μA full scale deflection. If a 100 μA meter is used, its effective sensitivity is increased more than ten times, to better than 10 μA full scale deflection.

When a voltage range multiplier *(R_{11})* is wired in series with the input lead as shown, the sensitivity of the resulting 'electronic voltmeter' is raised to 20,000 Ω/V using a 1 mA meter, and to 100,000 Ω/V using a 100 μA meter. Note that the circuit uses a stabilised supply of approximately 6 V, and this can be obtained via a zener diode.

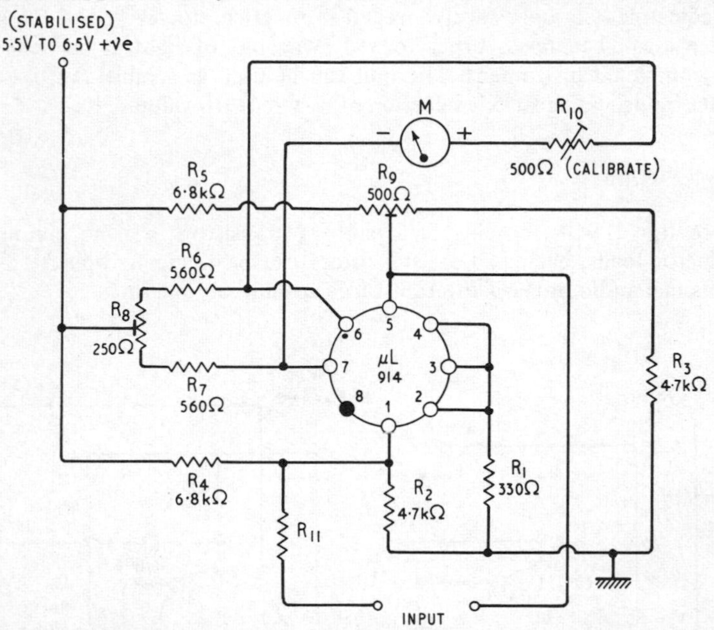

Fig. 5.24

D.C. meter amplifier/electronic voltmeter. R_{11} = *Voltage range multiplier*
= *20 kΩ/V for 1 mA meter*
= *100 kΩ/V for 100 μA meter*

Once the meter amplifier has been built, it must be set up as follows:

(1) Short the input terminals, and set R_8 for zero reading on the meter.

(2) Open the input terminals, and set R_9 for zero reading on the meter.

(3) Repeat (1) and (2) until no further zero adjustment is required.

(4) Select the multiplier *(R_{11})* for the full scale deflection voltage required, i.e., if a 1 mA meter is used and a full scale deflection fo 10 V is wanted, R_{11} needs a value of 200,000 Ω.

(5) Apply a known voltage that is close to the required full scale deflection to the input terminals, and adjust R_{10} to indicate this voltage on the meter. The unit is now ready for use.

When using the voltmeter, the above series of checks should occasionally be repeated, to make sure that calibration is correct. Re-adjustment is only rarely needed in practice, however, and R_8, R_9, and R_{10} can be pre-set types tucked away out of sight at the rear of the completed instrument. The unit can be used as a multi-range voltmeter by providing switch selection of different R_{11} values.

Mixer circuits

Since the μL914 contains two pairs of transistors sharing common collector loads, each pair of transistors can be made to operate as a 2-channel audio mixer if the transistors are suitably biased.

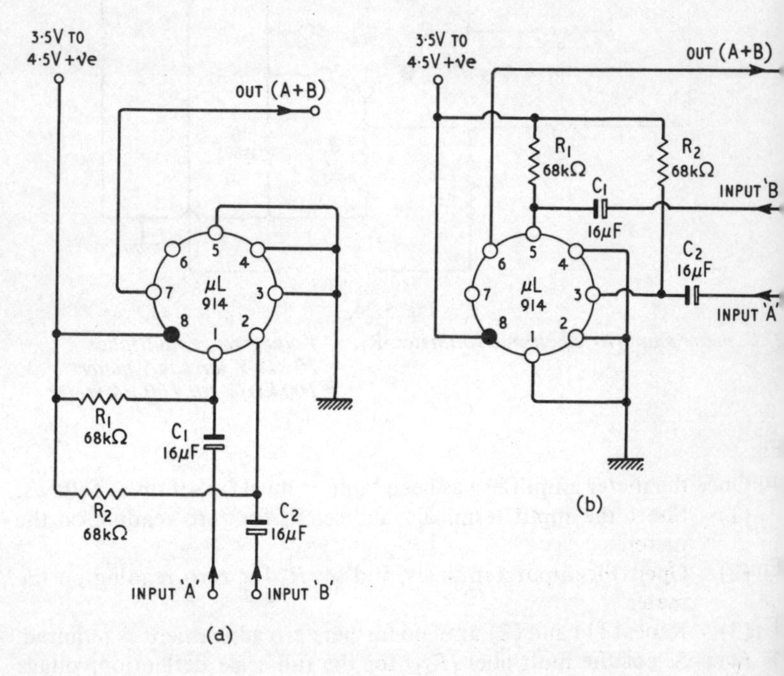

Fig. 5.25

(a) 2-channel mixer. (b) Alternative 2-channel mixer

Fig. 5.25a shows how to wire the i.c. as a 2-channel mixer, using the *Q*1-*Q*2 pair of transistors only, and Fig. 5.25b shows the alternative connection using the *Q*3-*Q*4 pair of transistors. Both of these circuits give a voltage gain of about sixteen times between each input and the output.

The i.c. can be made to serve as a 4-channel mixer by using both halves of the circuit and shorting all collectors to a common load.

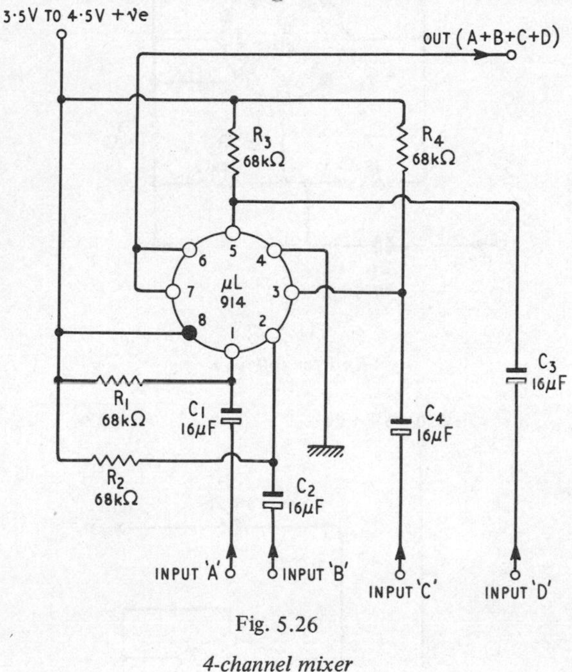

Fig. 5.26

4-channel mixer

Fig. 5.26 shows the circuit connections. The voltage gain between each input and the output is about eight times in this case.

Emitter follower

The i.c. can be used as a normal emitter follower by using the connections shown in Fig. 5.27. Input impedance is about 12 kΩ, but can be increased by using normal bootstrap techniques, if required. Only a single emitter follower can be made from each μL914.

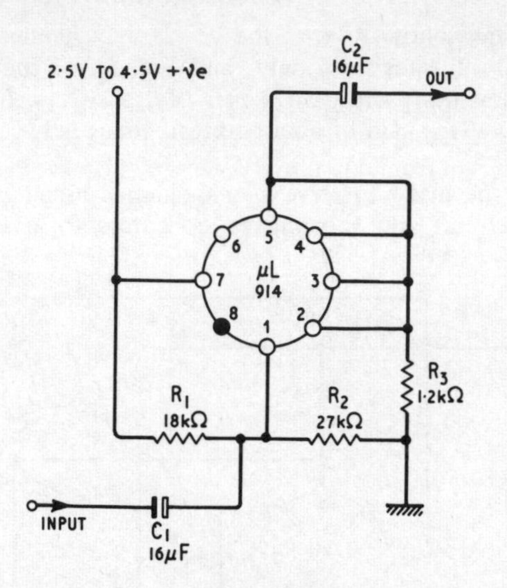

Fig. 5.27
Emitter follower

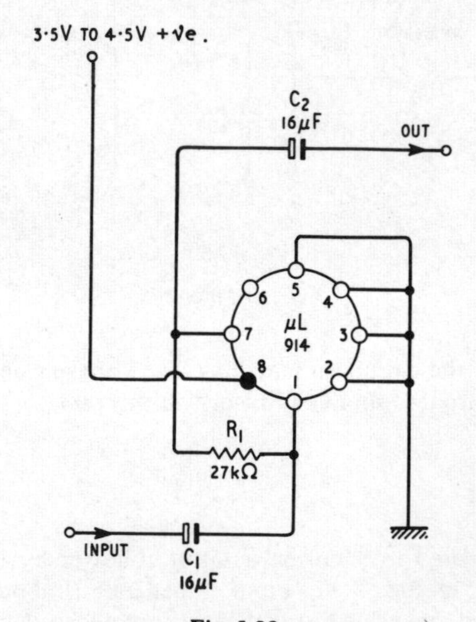

Fig. 5.28
Common emitter amplifier

Common emitter amplifiers

The differential amplifier of Fig. 5.21 can be used with advantage to replace a conventional common emitter amplifier stage, since it takes a near-constant current from the supply under actual operating conditions,

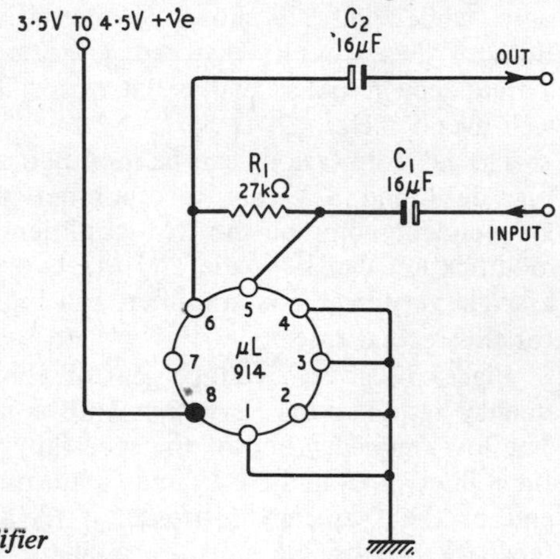

Fig. 5.29
Alternative common emitter amplifier

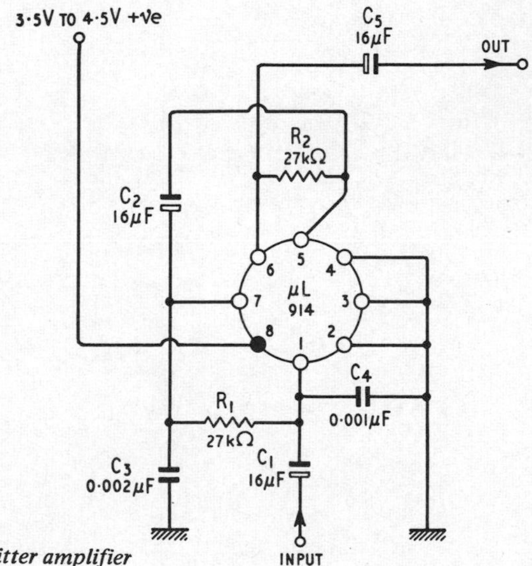

Fig. 5.30
Very-high-gain common emitter amplifier

and thus minimises the need for supply decoupling networks between the individual stages of a multi-stage amplifier. Only one of the Fig. 5.21 outputs is used in this type of application.

The μL914 can, however, be used as a conventional common emitter amplifier, if required, and Fig. 5.28 shows one connection using *Q*1 only, while Fig. 5.29 shows an alternative connection using *Q*4 only. Both of these circuits give voltage gains of about 20 times, and have a frequency response that is flat within 3 dB over the range 60 Hz to well over 1 MHz.

The i.c. connections can be modified so that the amplifiers of both Fig. 5.28 and 5.29 are incorporated in a single unit, making two independent common emitter amplifiers available. Alternatively, this modification can be made and the two amplifiers cascaded, to form a single very high gain amplifier, and Fig. 5.30 shows the connections for this second case.

Fig. 5.30 gives a voltage gain of about 400 times, and has a frequency response that is within 3 dB from 100 Hz to about 200 kHz. The low frequency end of the spectrum can be extended by increasing the values of C_1 and C_2. C_3 and C_4 are incorporated to limit the upper end of the frequency response. If C_3 and C_4 are removed, internal feedback causes the unit to oscillate violently at a frequency of several megahertz.

Index